Physical Methods for Microorganisms Detection

Editor

Wilfred H. Nelson

Professor
Department of Chemistry
Pastore Chemical Laboratory
University of Rhode Island
Kingston, Rhode Island

CRC Press
Taylor & Francis Group
Boca Raton London New York

CRC Press is an imprint of the
Taylor & Francis Group, an **informa** business

First published 1991 by CRC Press
Taylor & Francis Group
6000 Broken Sound Parkway NW, Suite 300
Boca Raton, FL 33487-2742

Reissued 2018 by CRC Press

© 1991 by CRC Press, Inc.
CRC Press is an imprint of Taylor & Francis Group, an Informa business

No claim to original U.S. Government works

Library of Congress Cataloging-in-Publication Data

Physical methods for microorganisms detection / editor, Wilfred H. Nelson.
   p. cm.
   Includes bibliographical references and index.
   ISBN 0-8493-4140-X
   1. Bioluminescence assay. 2. Microorganisms--Detection.
I. Nelson, Wilfred H., 1936-
QR69.B53P49   1991
576--dc20

91-2268

A Library of Congress record exists under LC control number: 91002268

Publisher's Note
The publisher has gone to great lengths to ensure the quality of this reprint but points out that some imperfections in the original copies may be apparent.

Disclaimer
The publisher has made every effort to trace copyright holders and welcomes correspondence from those they have been unable to contact.

ISBN 13: 978-1-315-89649-6 (hbk)
ISBN 13: 978-1-351-07559-6 (ebk)

Visit the Taylor & Francis Web site at http://www.taylorandfrancis.com and the
CRC Press Web site at http://www.crcpress.com

THE EDITOR

Wilfred H. Nelson has served as Professor of Chemistry at the University of Rhode Island, Kingston, since 1977. He received his B.Sc. from the University of Chicago in 1959 and his Ph.D. in 1963 from the University of Minnesota under the tutelage of R. S. Tobias. As a student he was an N.S.F. DuPont Fellow and a member of Phi Lambda Upsilon. In 1964, after completion of a postdoctoral year at the University of Illinois with D. F. Martin, he joined the faculty of the University of Rhode Island as Assistant Professor.

Dr. Nelson has been the recipient of numerous research grants. Awards have been received from the N.S.F., Research Corporation, U.S. Department of the Interior and the Petroleum Research Fund as well as the U.S. Army Research Office. He has published a large number of articles on topics ranging from Light-Scattering Theory and applications of the Kerr Effect to Analytical Microbiology. These have appeared in *The Journal of the American Chemical Society, The Journal of Physical Chemistry, Inorganic Chemistry, The Journal of Inorganic and Nuclear Chemistry, The Journal of Organometallic Chemistry* and *The Journal of Molecular Structure* as well as *The Journal of Microbiological Methods, Applied Spectroscopy, Applied Spectroscopy Reviews* and *American Laboratory*.

He has been an active member of The American Chemical Society, The Society for Applied Spectroscopy, The Society for Applied Bacteriology, the Biophysical Society and Sigma Xi. He is frequently invited to speak before both chemists and microbiologists. He has served as editor of books which have reviewed analytical methods applied to rapid microbiological analysis. His current research interests include the rapid detection of microorganisms, especially bacteria, with emphasis on the use of fluorescence lifetime and ultraviolet resonance Raman spectroscopies. He also studies the basic spectral properties of bacterial taxonomic markers including their electronic excited states.

ADVISORY BOARD
MEMBERS

CONTRIBUTORS

Fritz Berthold
Professor
Berthold Laboratorium
Wilbad, Germany

M. F. Chaplin
Department of Biotechnology
South Bank Polytechnic
London, England

Jillian S. Deans
Biosafety Group
Biological Treatment Division
Department of Trade and Industry
Warren Spring Laboratory
Stevenage, England

Charles T. Gregg
President
Bethco, Inc.
Los Alamos, New Mexico

M. W. Griffiths
Department of Food Science
University of Guelph
Guelph, Ontario, Canada

Brian J. McCarthy
British Textile Technology Group
Leeds, England

Robert Miller
Microbiology Laboratory
Tennent Caledonian Breweries, Ltd.
Wellpark Brewery
Glasgow, Scotland

Wilfred H. Nelson
Professor
Department of Chemistry
Pastore Chemical Lab
University of Rhode Island
Kingston, Rhode Island

Ole Olsen
Foss Electric
Hillerod, Denmark

T. T. Salusbury
Business and Environment Unit
Department of Trade and Industry
London, England

I. W. Stewart
Biosafety Group
Biological Treatment Division
Department of Trade and Industry
Warren Spring Laboratory
Stevenage, England

TABLE OF CONTENTS

Chapter 1
Bioluminescence in Clinical Microbiology 1
Charles T. Gregg

Chapter 2
Rapid Estimation of Microbial Numbers in Dairy Products Using ATP Technology 29
M. W. Griffiths

Chapter 3
Rapid Food Microbiology: Application of Bioluminescence in the Dairy and Food Industry
— A Review 63
Ole Olsen

Chapter 4
The Preparation and Properties of Immobilized Firefly Luciferase for Use in the Detection
of Microorganisms 81
M. F. Chaplin

Chapter 5
Bioluminescence Applications in Brewing.. 97
Robert Miller

Chapter 6
Rapid Enumeration of Airborne Microorganisms by Bioluminescent ATP Assay....... 111
Jillian S. Deans, I. W. Stewart, and T. T. Salusbury

Chapter 7
The Use of ATP Measurements in Biodeterioration Studies 129
Brian J. McCarthy

Chapter 8
Luminescence Instrumentation for Microbiology 139
Fritz Berthold

Index 151

Chapter 1

BIOLUMINESCENCE IN CLINICAL MICROBIOLOGY

Charles T. Gregg

TABLE OF CONTENTS

I.   Introduction... 2
      A.   What is Bioluminescence?... 2
      B.   Mechanism Studies, Properties of Bioluminescence 2

II.   Bioluminescence as a Rapid Screen for Urinary Tract Infection (UTI)........... 4
      A.   Early Work... 4
      B.   The Sociology of UTI... 4
      C.   Techniques for Urine Screening 5
      D.   Criteria for Evaluation of Urine Screens........................... 6
      E.   Future Developments 9

III. Bioluminescence for Rapid Antibiotic Susceptibility Testing (AST) 10
      A.   Introduction 10
      B.   Early Work... 10
      C.   Current Standard Methods... 10
      D.   The Bioluminescence AST 11
      E.   The LAD Sensi-Quik™.. 12
            1.   Introduction... 12
            2.   Medium Studies... 13
            3.   Ionic Strength Effects....................................... 13
            4.   Incubation Times 14
            5.   Antibiotic Concentration Curves 14
            6.   Error Rates.. 14
            7.   Microtiter Plate (MTP) Luminometer Evaluations 17
            8.   Current Status... 20

IV.   Concluding Remarks... 22

Acknowledgments... 23

References.. 24

Then shall thy light spring forth as the morning, and thine health shall spring forth speedily.

Isaiah 58:8

I. INTRODUCTION

A. WHAT IS BIOLUMINESCENCE?

The quote from Isaiah succinctly summarizes the concerns of this chapter, i.e., light (from bioluminescence) and its role in health care.

Luminescence is a term coined in 1888 by the physicist Eilert Weidemann to distinguish "cold" light from the light arising from incandescence.[1,2] Weidemann indicated the source of energy for the varieties of cold light by various prefixes. In chemiluminescence, for example, the source of energy is a chemical reaction.

Bioluminescence is a subcategory of chemiluminescence in which the energy-generating chemical reaction is enzyme catalyzed; enzymes are produced only by living systems.

Bioluminescence was described by early Chinese and Greek writers (e.g., Aristotle — *De Anima*), but little understanding of the process was achieved until late in the nineteenth century, although Robert Boyle had shown two centuries earlier that the process required oxygen, thereby implying a chemical reaction.[3] In 1885, Dubois noted that a cold water extract of the luminous organ of a shellfish, although luminous initially, gradually darkened. Luminosity could be restored if a hot water extract of the same organ was added to the now inactive cold water extract.[4] Dubois reasoned that an enzyme in the cold water extract consumed a substrate in the light-generating reaction. This substrate was present in the hot water extract, and adding it made further light generation possible until it too was consumed. He called the enzyme luciferase, and the substrate luciferin, from the Greek word (lucifer) for light-bearer. These are the names we have used ever since.

At the Paris International Exposition of 1900, Dubois exhibited a 24-l jar of "Photobacteria" and nutrients that gave enough light for reading newspapers in an otherwise darkened room.[5]

One form of bioluminescence is the source of the light from the luminous beetles we call fireflies. Light generation in the firefly is stoichiometric with adenosine triphosphate (ATP) consumption with a quantum yield of 0.88.[5] By contrast, the quantum yield from nonenzymic chemiluminescence reactions is seldom more than 0.01.[6] Bacterial luciferase also produces light, but the chemistry of the reaction is quite different. In this case, light production depends upon the oxidation of reduced pyridine and flavin mononucleotides by oxygen in the presence of a long chain aldehyde.

Although both chemiluminescence and bacterial bioluminescence have important applications, particularly in clinical chemistry and in research, I will concentrate here on the ATP-dependent light generation from firefly luciferase, and on its applications to clinical microbiology. This topic has been reviewed in recent years by many well-qualified investigators whose efforts have greatly contributed to my own understanding of this fascinating field.[6-11]

B. MECHANISM STUDIES, PROPERTIES OF BIOLUMINESCENCE

It was more than a half century after Dubois that Professor William D. McElroy and his students at Johns Hopkins University began, in 1942, by repeating Dubois's experiments, then evolved a description of the chemistry of the firefly system that was nearly complete some 40 years later.[12]

Forty-seven years after he began his research in bioluminescence the redoubtable Professor McElroy, at this writing, remains active in highly original research.[13]

The reaction of light generation by the firefly system may be written:

$$\text{luciferin} + \text{luciferase} + \text{ATP} + \text{Mg}^{2+} \longrightarrow$$
$$\text{luciferyladenylate-luciferase} + \text{pyrophosphate}$$
$$\text{luciferyladenylate-luciferase} + \text{O}_2 \longrightarrow$$
$$\text{luciferase} + \text{oxyluciferin} + \text{AMP} + \text{light} + \text{CO}_2$$

ATP is nearly always used as an energy source in metabolic reactions; the luciferin/luciferase reaction is an exception. Here ATP functions to link the enzyme luciferase to its substrate luciferin. The energy driving the reaction comes, as in luminescence reactions generally, from the oxidation in step 2.

Nevertheless, ATP can be measured in this system with exquisite sensitivity. Measurement of 1.0 fmole (0.5 pg) (femto [f] $= 10^{-15}$ pico [p] $= 10^{-12}$) is routine; with care 50 fg of ATP can be measured.[5,14] This is approximately 10 million times less ATP than can be measured by standard fluorimetry.[5]

Even greater sensitivities have been reported. Stanley suggests that the best instruments can detect 10 fg of ATP, or 10^7 molecules.[15] Campbell and associates, using specially prepared reagents and a new luminometer (the CLEAR SpeedTech 2000) report detection of 25 bacteria (about 25 fg of ATP, see below).[16] An independent evaluation of the CLEAR luminometer reports a sensitivity only slightly better than the LAD 535 luminometer or the Turner model 20.[17] This report must be considered preliminary, however, because of the conditions under which the CLEAR instrument was tested.

The luciferase reaction is highly specific for ATP.[18] The next most effective substrate is deoxy-ATP, which is 1.7% as active as ATP; 11 other nucleoside triphosphates are less than 0.1% as effective as ATP.

Since ATP is a component of all forms of life on earth, and is rapidly destroyed when a cell dies, the extraordinary sensitivity of the firefly system is an obvious choice for detecting very small numbers of living cells in a variety of environments. Currently, it is employed in commercial systems in clinical microbiology for screening for urinary tract infection; commercial systems for determining the antibiotic susceptibility of infecting organisms are in early clinical trials.

Following the pioneering work of McElroy and his colleagues,[19-22] these applications were utilized in early work at NASA on clinical applications as well as determining the existence of life on other planets.[23-26] These studies are the basis of the bioluminescence tests for bacteriuria and antimicrobial susceptibility as used today.[24,26] Measurement of ATP in the presence of an effective antibiotic is also a way of determining the presence of any noncellular ATP.[27]

The detection limits for bacteria by the firefly luciferin/luciferase assay clearly depend upon the amount of ATP in a single bacterium, usually a colony forming unit, or cfu. The concept of a cfu, however, is not as simple as it first appears.

For the Gram negative rods that make up the large and important genus *Enterobacteriacae,* a cfu is usually a single organism, at least in the absence of conditions that promote clumping. For the clinically important Gram positive cocci (*Staphylococci* and *Streptococci*), however, ATP estimation of cell numbers by bioluminescence may not give good agreement with the colony count on an agar plate.

Staphylococci and *Streptococci* grow in bunches, and chains, respectively. Each cfu of these organisms may, in fact, represent 10 to 20 bacteria. Bioluminescence measurement of ATP counts the contribution from each cell, thus, under some circumstances, giving an apparent false positive. For example, the ATP method may report 10^5 cfu/ml (of individual bacteria) while the colony count shows 10^4 cfu/ml, or less.

The concentration of ATP per bacterium also varies by at least an order of magnitude among various species of bacteria. It also varies for a particular bacterial population as the population moves from lag phase to exponential growth, and thence to stationary phase. In

TABLE 1
Urinary Tract Infection Syndromes

	Asymptomatic bacteriuria	Catheter-associated bacteriuria	Cystitis	Acute pyelonephritis	Acute urethral syndrome
Bacterial #	$\geqslant 10^5$ cfu/ml	$\geqslant 10^5$?	$\geqslant 10^5$	$\geqslant 10^5$	ca. 10^2
Urgency, frequency, dysuria	−	−	+	−	+
Chills, vomiting, fever $\geqslant 38°C$	−	−	−	+	+
Pyuria	±	±	+	+	±
Gram stain	+/−	+	+	+/−	+/−

addition, the ATP content of a bacterial population may vary depending on the conditions of their isolation and the time before measurement. A further confounding factor is the presence in the bacterial milieu of substances that inhibit firefly luciferase.

Careful investigation indicates that the ATP content of an "average" bacterium is at least 0.25 to 0.47 fg, with a range of more than tenfold.[27-29] Thus, the firefly system is potentially capable of detecting at least 100 to 1000 bacteria under optimal circumstances. Yeast cells, or red blood cells contain approximately 100 times as much ATP as bacterial cells, while other mammalian cells are equivalent in ATP content to about 1000 bacteria.

II. BIOLUMINESCENCE AS A RAPID SCREEN FOR URINARY TRACT INFECTION (UTI)

A. EARLY WORK

The first paper in the new *Journal of Clinical Microbiology* in 1975 was on the use of bioluminescence as a urine screening method.[30] This was almost simultaneous with a paper by Conn et al.[31] on the same subject, and was followed shortly by a report by Johnston et al.[32] There was a comparative lull in publications in this area until 1982,[33,34] and for 2 years after that. It was the calm before the storm. More papers on bioluminescence screening for urinary tract infection appeared in 1984 to 1985 than were published in the preceding decade.[35-45] The results of some of these investigations will be discussed later (Table 2).

B. THE SOCIOLOGY OF UTI

The exploding interest in this application is not surprising. Urine is the largest single category of specimens received in the clinical microbiology laboratory, making up 30 to 50% of the total; 50 to 80% of these specimens are negative, so the efficient screening out of these negative specimens can be highly cost effective for the laboratory.

Community-acquired UTI account for 6 to 7 million doctor's office visits in the U.S. each year.[46] Approximately a quarter million of these infections lead to hospitalization and i.v. antibiotic therapy. In about one fifth of UTI victims the disease recurs three to six times a year.[46] One or two of every ten women will have at least one community-acquired UTI in her lifetime.

About one million hospitalized patients will suffer a nosocomial (hospital-acquired) UTI each year.[47] The disease may be fatal in the elderly, and is a precursor of up to half of Gram negative bacteremias (a life-threatening infection of the blood). UTI is the most common complication of pregnancy, and it is particularly serious.[47] Pregnant women with UTI have 26 times the incidence of pyelonephritis (kidney infection), compared to non-pregnant women; there is a 70% increase in low birth weights, and a 60% increase in prenatal deaths following UTI in pregnancy.[48]

Table 1 summarizes the parameters of the various UTIs.

Table 1 shows that although a cutoff of 10^5 cfu/ml is adequate for the majority of patients with UTI, it is inadequate to diagnosis of acute urethral syndrome and is being questioned in cases of catheter-associated bacteriuria. Fortunately, the symptoms of acute urethral syndrome are sufficiently compelling that a physician would probably begin antibiotic therapy without waiting for the results of urine culture or screening.

In traditional clinical microbiology laboratories, all urine specimens are subjected to the century-old technique of plating on one or more agar plates, incubating for 18 to 24 h, and then examining the plates for bacterial colonies. Since the majority of urine specimens are negative, a considerable amount of time and material is consumed by negative specimens. To reduce this wasted effort, particularly in this era of stringent control of treatment costs, many clinical microbiology laboratories have been forced to screen urines to identify positive specimens quickly. This means that further effort can be concentrated on the approximately 20% of total urine specimens that are positive.

C. TECHNIQUES FOR URINE SCREENING

A number of urine screening techniques have been used. Those most commonly employed at present include the Chemstrip LN (for leukocyte-nitrite) dipstick, the Bac-T-Screen, the Gram stain, and the Los Alamos Diagnostics (LAD) bioluminescence UTIscreen® kit. Evaluations of all these screening methods have been recently published.[49-51]

The Chemstrip LN (Boehringer Mannheim Diagnostics, Indianapolis, IN), is a plastic dipstick that detects leukocytes by measuring leukocyte esterase. The enzyme, if present in urine, hydrolyzes a synthetic substrate leading to a purple color on the dipstick. The reduction of nitrite by bacterial enzymes is also measured. The resulting nitrate gives a pink color on the dipstick.

For the most accurate results, the LN strip should be used only for urines that have incubated in the bladder for at least 4 h (e.g., morning urines). Leukocytes are routinely present in urine in significant numbers in pyelonephritis and cystitis and sometimes in other lower UTIs as well (Table 1). The nitrite reductase test works only for Gram negative organisms in the absence of ascorbate (Vitamin C), or other interfering substances (e.g., nitrate from the diet). The urine must also have a pH greater than 6.0[52] Use of the dipstick is a simple manual procedure.

Bac-T-Screen (Vitek Systems, St. Louis, MO) is a system in which urine is filtered, the filter washed and stained, and the amount of dye retained compared either visually or photometrically to a standard. Bac-T-Screen is available as a semi-automated instrument in which the technologist simply adds the specimen and then replaces the filter after reading.

The Gram stain is a well-known and effective microscopic method; it will not be described here.

The LAD UTIscreen measures the bacterial ATP content of urine. In patients with UTI, the urine may contain ATP both from bacteria and from somatic cells (erythrocytes, leukocytes, and the epithelial cells lining the urinary tract). Urine may, on occasion, be visibly bloody. In this case the urine contains more than about 10% blood by volume, and may thus contain 10^8 red blood cells per ml. In pyelonephritis the urine may contain up to 10^7 leukocytes per ml.

In the UTIscreen procedure, 25 μl of urine are added to a tube containing lyophilized detergent and potato apyrase, an enzyme that hydrolyzes ATP. The urine is incubated in this tube at room temperature for 10 to 40 min, at the convenience of the technologist running the test. Incubation for less than 10 min can lead to excessive levels of false positive results, while incubation for more than 40 min can lead to release and hydrolysis of the bacterial ATP, with resulting false negatives.

The amount of apyrase in the UTIscreen tubes can hydrolyze the ATP liberated from about 10^6 somatic cells per ml of urine, whether they are leukocytes or epithelial cells. If the urine is visibly bloody (i.e., if it contains more than 10^8 red blood cells per ml), however, or if it contains more than 10^6 other somatic cells per ml, some somatic cell ATP will remain after a 10 min incubation, and the test may read falsely positive. A longer incubation will solve this problem.

Following the incubation with apyrase, the UTIscreen tube is placed in an LAD lumi-nometer (either the semi-automated model 535, or the fully automated model 633). A detergent mixture (called BRA, or bacterial releasing agent), is then automatically added by either of the LAD luminometers to release the bacterial ATP, a luciferin/luciferase mixture is added automatically, the light is measured after a 5 s delay, and the result is displayed and simultaneously printed.

D. CRITERIA FOR EVALUATION OF URINE SCREENS

A number of criteria enter into the evaluation of a urine screening test. These include performance parameters such as sensitivity, specificity, the predictive value of a negative test, and test efficiency.[53] Cost is also an important factor.

Sensitivity, specificity, the predictive value of a negative test, and test efficiency were defined by Galen and Gambino as follows.[53]

$$\text{Sensitivity} = \text{TP}/(\text{TP} + \text{FN}) \times 100$$
$$\text{Specificity} = \text{TN}/(\text{TN} + \text{FP}) \times 100$$
$$\text{Predictive Value of a Negative Test (PVN)} = \text{TN}/(\text{TN} + \text{FN}) \times 100$$
$$\text{Efficiency} = \text{TP} + \text{TN}/(\# \text{ of specimens}) \times 100$$

Where TP = true positives, TN = true negatives, FP = false positives, and FN = false negatives.

Sensitivity is a measure of a test's ability to identify a specimen from a diseased patient correctly (i.e., as a true positive), while specificity measures the test's ability to identify a specimen from a healthy subject as a true negative. The PVN indicates the extent to which a negative result is truly negative. Efficiency indicates the accuracy of the test, i.e., the percentage of the total measurements that are correct.

Predictive values vary with the prevalence of disease in the population being studied.[53] For example, with a test of sensitivity of 96% and a specificity of 70% the Predictive Value of a Positive test (PVP) is 38% if the incidence of UTI in the population under study is 15%, while if the incidence of UTI is 50% (as in some nursing homes and catheterized patients), the PVP rises to 89%. Conversely, the PVN drops from 97 to 82%. The user must beware of extrapolating predictive values such as those in Table 2 for essentially normal populations to a selected population in which the incidence of disease is much different from 10 to 20%.

As with all statistical measurements, an adequate number of specimens must be run when evaluating a test. Usually, at least several hundred specimens need to be run before meaningful conclusions can be drawn.

False positives have two principal effects. They may mean that a healthy patient is given unneeded antibiotic therapy until another specimen is taken (usually the next day). A false positive also means that the laboratory may unnecessarily plate that specimen. Plating substantially increases the cost per specimen (see below).

A false negative may result in delay in recognizing the presence of UTI, although the physician may elect to begin treatment based on his or her clinical judgement. Whether we as laboratorians like to admit it or not, laboratory tests, whether from clinical chemistry or clinical microbiology, play a very modest role in the physician's decisions about treatment. One estimate is that patient history contributes about 70%, physical examination of the patient contributes 20%, and laboratory tests add another 10% to the decision-making process.[54]

Table 2 shows a comparison of UTIscreen™ sensitivity, specificity, and the predictive value of a negative test (PVN), as determined by Pezzlo et al.[50] with other bioluminescence urine screening studies at $\geq 10^5$ cfu/ml, as well as a group of five studies conducted with the LAD model 633 luminometer at $\geq 10^4$ cfu/ml compared to other studies.

TABLE 2

Comparison of UTIscreen™ with other Bioluminescence Urine Screening Studies

Study	Year	# Specimens	Method	Sens.	Spec.	PVN	Ref.
AT $>10^5$ cfu/ml							
Pezzlo	1989	1000	LAD 535[a]	96%	70%	98%	50
Schifman	1984	973	Turner	89	63	95	39
Welch	1984	986	Turner	91	63	97	40
Welch	1984	986	Lumac	95	76	98	40
Kolbeck	1985	1000	Turner	95	92	98	42
Wu	1985	2815	Turner	90	24	57	45
AT $>10^4$ cfu/ml							
LAD[b]	1989	5099	LAD 633[a]	85	63	87	
Drow	1984	986	Lumac	92	79	95	36
Drow	1984	986	Monolite	89	82	96	36
Bixler	1985	848	Lumac	94	67	98	41
Park	1984	2000	Lumac	96	71	99	38

[a] The LAD model 535 is a semi-automated luminometer capable of running 60 specimens per hour, the model 633 is fully automated, and will run 150 specimens per hour. The Turner, Lumac, and Monolite bioluminescence urine screening kits are no longer available.

[b] Unpublished data.

With the LAD luminometers an ATP standard is run to insure that both instrument and reagents are working properly, and to establish a reference value for the light output. LAD recommends at the outset that a value of 10% of the light output from the ATP standard be taken as the working cut-off point, i.e., the boundary between positive and negative specimens.

With further experience, the director of the clinical microbiology laboratory and his or her clinical colleagues can establish a new cut-off point that best suits their patient population, specimen handling, and their desire to minimize false positives at the expense of increasing false negatives, or the other way around.

Decreasing the cut-off value increases sensitivity, the false positive rate, and the PVN, while decreasing specificity, the false negative rate, and the PVP.

In Table 2, Pezzlo et al.[50] used a 5% cut-off, while in the tests at 10^4 cfu/ml the five laboratories involved found that a 2% cut-off gave them the most satisfactory results.

The LAD model 535 UTIscreen™ system (instrument and reagents) gave values comparable to those found by other workers using other systems (Table 2). It is noteworthy that the range of values obtained by various investigators all using the same instrument, reagents, and analytical procedure was substantial.

The LAD model 633 UTIscreen™ system gave acceptable results at 10^4 cfu/ml, but the other systems performed somewhat better.

Table 3 shows the values of sensitivity, specificity, and the predictive value of a negative test obtained by Pezzlo et al.[50] in a 1000 specimen evaluation of UTIscreen, Bac-T-Screen, the LN strip, and the Gram stain. The specificity of UTIscreen™ is substantially higher than either the LN strip or the Bac-T-Screen, and the PVN is higher as well. The values of these parameters for UTIscreen were identical to those of the Gram stain.[51]

A major criterion for the evaluation of a urine screening test is cost, mainly because of the large numbers of specimens to be processed. Unfortunately, this is an area in which there is much misunderstanding. The "cost" of a test is frequently said to be the cost of the supplies needed to perform it. This is like saying that the cost of driving an automobile is approximately $0.06/mi for gasoline, when operating even a compact car costs over $0.40/mi.

TABLE 3
Comparison of LAD UTIscreen™ with other Current Urine Screening Tests (%)

	Sensitivity	Specificity	FP	FN	PVN	Efficiency
UTIscreen—5% cut-off	96	70	23	1.2	98	86
LN strips	90	51	40	3.1	93	69
BAC-T-Screen	96	48	42	1.2	97	66

Adapted from Pezzlo et al., *J. Clin. Microbiol.*, 27, 716, 1989.

In nearly all tests a major component of the test cost is labor. A major additional cost of urine screening is the added expense of plating specimens that the test falsely identified as positive. The result is that the actual cost of urine screening rises abruptly as the percentage of false positives for the various urine screening methods increases.

An example of this misunderstanding is in the cost of plating a urine specimen on two or more types of agar. Based on U.S. costs in 1989, the "true" cost of plating is $1.60 to 1.63.[4,50] This includes a materials cost of about $0.60, and a labor cost of about $1.00. The labor cost, in turn, is based on a figure of 5 min per plate (at a technologist's salary of $12.00/h). Five minutes is the time provided for plating from the American College of Physicians (CAP) time and motion studies in a number of laboratories. Although the CAP "workload units" have been recently criticized,[55] the value of 5 min for plating has been independently confirmed.[56]

Even the figure of $1.63 for plating is falsely low. The fully burdened cost of a technologist (including fringe benefits, overhead, etc.) is approximately $30.00/h.[56] If this cost is used in place of the salary figure, the cost of plating is about $3.13 per specimen.

Pezzlo et al.[50] calculated the cost per test for plating, UTIscreen™, Bac-T-Screen, Chemstrip LN, and the Gram stain. The costs included materials, time required to perform the test (at $13.00/h for technologist's time), and the cost of plating false positives. Since the false positive rates of both Bac-T-Screen and Chemstrip LN are high, their costs per test are correspondingly elevated.

The resulting costs per urine specimen found by Pezzlo et al.[50] were

plating	=	$1.63
UTIscreen™	=	1.75
Chemstrip LN	=	1.43
Gram stain	=	1.97
Bac-T-Screen	=	2.12

Pezzlo et al.[50] used a time of 4 min for plating. If 5 min were used instead it would increase the cost of plating to $1.85 per specimen for materials and labor alone.

The Gram stain, although a very useful test, is tedious, time consuming, and consequently, relatively expensive as a urine screening test.

Although the Chemstrip LN is simple to perform and slightly less expensive than the other tests it is too inaccurate to be used alone for urine screening.[57,58]

False positives in UTIscreen™ can arise from a number of sources, some of which have already been discussed (the presence of somatic cells at $>10^6$/ml, for example, if a 10 min incubation is used, and clusters or chains of *Staphylococci* or *Streptococci*, respectively).

An important source of false positives is the presence in urine of fastidious or slow-growing organisms that do not grow up to form visible colonies overnight on McConkey's or blood agar. These organisms include diptheroids, lactobacilli, coagulase negative *Staph-*

ylococci, viridans *Streptococci,* anaerobes, viruses, protozoa, and yeasts.[59] The contribution from this source can be evaluated by keeping plates for 48 h before concluding that they are negative.

False positives can also arise in a bioluminescence method if either the reagents or the reagent-dispensing lines become contaminated. Rigorous cleaning of the lines should solve the first problem, and careful storage of the reagents will solve the second. Light leaks into the instrument (usually at the point where the reagent-dispensing lines enter the sample chamber), can also cause false positives. Both this source of false positives and that due to contamination of reagents or dispensing lines are indicated by the presence of high blank readings.

False negatives may arise from the presence in the urine of strong inhibitors of luciferase, usually in the form of high ionic strength. Antibiotic treatment of the patient's infection might be expected to cause a problem but apparently does not.[50]

Enterococci and yeast each make up about 8% of normal urine specimens.[50] UTIscreen® missed about 14% of both, for a false positive rate for each genus of slightly more than 1%. These organisms are somewhat resistant to the action of the BRA used in UTIscreen. Better results should be obtained if the interval between the addition of BRA and luciferin/luciferase were increased to 5 s (as it is in the model 633), or even longer. Experiments are in progress to shed additional light on this problem.

Another cause of a series of false negatives is allowing the luciferin/luciferase bottle to run dry before the end of a batch of specimens. Since the bottle is amber (to protect luciferin from light), the liquid level can be difficult to see. The operator should be alert to how many specimens can be run before a new bottle of luciferin/luciferase must be substituted.

False negative rates are calculated here as the percentage of the total number of specimens that are false negatives, i.e., those that fail to exhibit enough light output in the bioluminescence assay to exceed the chosen cut-off value, but that give visible colonies on plating.

This way of measuring false negatives is independent of the prevalence of UTI in the population being studied. One thousand specimens that yield 20 false negatives have a false negative rate of 2%. If false negatives are considered as a percentage of the total positive specimens, however, then the false negative rate is a linear function of the prevalence of UTI. Twenty false negatives in 1000 specimens is a false negative rate of 20% if 100 specimens are positive (a 10% incidence of UTI), while the false negative rate is 4% if the incidence of UTI in the study population is 50%. Galen and Gambino[53] define false negatives as a percentage of true positives, but the definition of false negatives as a percentage of the total number of specimens is more commonly used.

E. FUTURE DEVELOPMENTS

Future developments will probably include use of immobilized luciferase on membranes,[60,61] on ion exchange columns,[62] in the form of a dipstick,[63] or as part of a flow-through system.[32,64] Immobilized luciferase is more active and substantially more stable than luciferase in solution.

Luciferin is not generally available as a completely pure compound, and the impurities include oxyluciferin and dehydroluciferin, both of which inhibit luciferase.[15,65] Better commercial preparations will enhance the sensitivity of the assay.

Agents used to extract ATP from bacteria generally inhibit luciferase. Better extractants,[63] or methods of protecting luciferase against the extractant inhibition,[66-68] or a combination will increase the sensitivity of the ATP assay.

Firefly luciferase produced in bacteria into which the firefly gene has been cloned is already a commercial reality, although its current price limits its utility. The cost of collecting fireflies has risen from 1¢ each in the early 1960s to only 1.3¢ nearly 30 years later. The American firefly is far from being an endangered species.

In addition to better reagents, more sensitive instruments may become available. These may replace the photomultiplier entirely with a different kind of detector.[69]

III. BIOLUMINESCENCE FOR RAPID ANTIBIOTIC SUSCEPTIBILITY TESTING (AST)

A. INTRODUCTION

The great variety of infectious bacteria, and their widely differing response to various antibiotics present the physician with a complex problem. Once he or she has determined that the patient has an infection the next question is, "How can I most effectively treat this patient?"

To phrase the question differently, what the physician generally needs to know is: to which antibiotics is the infecting organism susceptible? Susceptible means that the growth of the bacterium is stopped (the antibiotic is bacteriostatic), or the bacterium is killed (the antibiotic is bactericidal).

Since patients with infectious disease still occupy more hospital beds than those suffering from heart disease, cancer, and stroke combined, it is not surprising that information on the antibiotic susceptibility of bacteria (the Antibiotic Susceptibility Test, or AST), is the clinical microbiology test most frequently requested by physicians.

The classical, and still most widely used, forms of the AST require 18 to 24 h to produce results. In serious infections, the physician cannot wait that long to begin treatment. Typically, a broad-spectrum antibiotic is prescribed that will probably be effective against the infecting organism. Unfortunately, the broad-spectrum antibiotics are expensive (a course of treatment can cost several thousand dollars), and they are frequently toxic, i.e., they have side effects so severe that the antibiotic must be discontinued.

The bioluminescence AST developed by Los Alamos Diagnostics (Sensi-Quik™), evaluates the antibiotic susceptibility of infecting organisms in 3 h rather than 18 to 24 for more than 90% of clinically significant bacteria. This gives the physician the option of witholding treatment until he or she knows which antibiotics will be effective, or changing to a less expensive or less toxic antibiotic after only a few hours.

B. EARLY WORK

Antibiotic susceptibility testing began with Alexander Fleming's accidental discovery, in 1929, of the antibiotic produced by a mold colony (*Penicillium notatum*) on a culture dish containing the Gram positive pathogen *Staphylococcus aureus*.[70] There was no growth of the *Staphylococci* in the immediate vicinity of the mold. Fleming recognized the significance of this observation, and went on to prepare a crude extract of the mold and demonstrate its inhibitory action on many species of bacteria.

Penicillin was difficult to purify, however, and significant progress was not made in the development of susceptibility methods until the 1940s, after several more important antibiotics had been discovered and were being used by physicians.

The earliest method of testing involved putting antibiotic-impregnated paper strips on agar uniformly inoculated with bacteria, and observing the zones of inhibition around the paper when "susceptible" strains of bacteria were used. The agar was clear for some distance around the paper strips that contained an effective antibiotic. If a strip contained an antibiotic to which the organism was resistant, bacterial growth was present up to the edges of the strip.

C. CURRENT STANDARD METHODS

The paper strip method has been progressively refined.[71-73] In its present form, the antibiotics are contained in paper disks, the disks are placed on agar plates uniformly inoculated with bacteria, and, following an 18 to 24 h incubation at 35°C, the diameters of the zones of inhibition around each antibiotic disk are transformed into "category" results (Susceptible, Intermediate, or Resistant — SIR), for progressively smaller zones of inhibition. The difference between an Intermediate and either a Resistant or Susceptible organism is often only 1 mm so the zones of inhibition must be carefully measured with vernier calipers. This is known as the Kirby-Bauer method, although the authors of the original paper were Bauer, Kirby, Sherris, and Turck.[71]

Other methods that yield comparable results have also been developed. In the broth dilution method, increasing concentrations of an antibiotic are added to a series of tubes containing broth inoculated with a particular bacterium. After 18 to 24 h there is growth (turbidity) in all tubes up to a certain concentration of the antibiotic. Above that concentration the bacteria do not grow and the tubes are clear. The concentration of the antibiotic that just inhibits visible growth is called the Minimum Inhibitory Concentration, or MIC.

The Kirby-Bauer method gradually became the most common AST method because of the large number of tubes that had to be prepared in the broth dilution method. Now, however, the broth dilution method has been adapted to a microtiter plate format so that 96 "tubes" (microtiter wells) can be prepared and evaluated with comparatively little technologist time. The result is that broth dilution is currently the dominant method for determining antibiotic susceptibility.

Details of the standard methods for conducting and interpreting these tests are available from the National Committee on Clinical Laboratory Standards (NCCLS). NCCLS currently recommends that physicians be given only category results (SIR) even if an MIC was run, because the SIR results are more readily interpreted by physicians who are not infectious disease specialists.

The availability of AST data to clinicians has a significant effect on the management of hospitalized patients.[74,75] This information may bring about a change in therapy to less toxic or expensive antibiotics, or to more appropriate antibiotics that hasten recovery and reduce hospital stays. Since the average cost per hospital day in the U.S. is currently about $814[76] rapid AST data has substantial economic importance.

D. THE BIOLUMINESCENCE AST

In the bioluminescence AST test, bacteria are grown in broth in the presence or absence of antibiotics. Growth is indicated by an increase in the ATP level. At the end of the incubation period, the ATP is extracted with a detergent that makes the bacterial cell wall permeable to ATP, and the ATP is measured in the luciferin/luciferase system.

This procedure was developed by NASA in the mid-1970s as an extension of research designed to detect life on other planets,[25] although the idea of using the ATP bioluminescence to measure both bacteriuria and antimicrobial susceptibility had been the subject of an M.S. thesis in 1970.[77]

The value of this approach was quickly expanded by Swedish investigators.[78-81] Fauchere et al. also developed a bioluminescence susceptibility test in 1982,[82] and McWalter applied bioluminescence to the determination of the antibiotic susceptibility of S. aureus to methicillin in 1984.[83] He later reported results on other antibiotics.[84] McWalter's results are roughly comparable to our own, but his investigations were confined to a single bacterial species. Beckers and co-workers in Germany attempted to use bioluminescence to assess the effects of ampicillin on Escherichia coli, but concluded that the method was unsuitable.[85,86] Kouda et al. developed a similar assay in Japan in 1985.[87] Barton also used a bioluminescence method to examine methicillin resistance in S. aureus.[88] Preliminary experiments using the bioluminescence technique were reported at the Annual Meeting of the American Society for Microbiology in 1985. Schifman and Delduca used a group of Enterobacteriaceae and hypotonic medium with six antibiotics and a 3 h incubation. Their results gave good agreement with standard methods.[89] Park reported bioluminescence studies of Amphotericin B on yeast and got good agreement with standard methods using a 3 h incubation.[90]

In 1986 to 1987, the first results were reported from Hastings and his colleagues in Sheffield.[91] This very active group used an Amerlite microtiter plate luminometer and LKB reagents, a 1 h (or longer) preincubation at 37°C, followed by incubation of forty strains of *Enterobacteriacae* (including ten strains of *Proteus mirabilis*), with ampicillin or gentamicin for 3 h at 37°C. ATP was extracted with trichloracetic acid (TCA) and EDTA at 80°C. With *P. mirabilis*, a hypotonic medium was required to give satisfactory results with a standard method.

This work was later extended to encompass over 300 clinical isolates of both Gram positive and Gram negative organisms. Agreement with the reference method was 94% or better.[92] In a subsequent paper, the Sheffield group reported that hypotonic medium was used only with *Proteus*, experiments on *Staphylococcus* were run at 30°C, and the incubation period had to be extended when *Pseudomonas aeruginosa* was run against beta-lactam antibiotics (the penicillins, cephalosporins, and imipenem).[93]

Although basically similar, the assays used by these various groups differ considerably in detail. Most used specimens from agar plates (sometimes with a preincubation phase of 1 to 3 h), but overnight broth cultures were used by some groups.[11] Inoculum sizes varied over three orders of magnitude, and either intracellular ATP, total ATP, extracellular ATP, or some combination was measured.

Franc reported from Prague on a bioluminescence susceptibility test applied directly to positive urine specimens. Only gentamicin was studied; the test was run for 2 h at 37°C.[94] The Swedish group reported further results on both bacteria and fungi.[95,96] In this and earlier studies, they had used both beta-lactam antibiotics (e.g., ampicillin), and aminoglycosides (e.g., gentamicin), against bacteria, and two antifungal agents against yeast. With ampicillin and either *P. mirabilis* or *E. coli* they got clear-cut differences after 1 to 2 h. This group employed boiling Tris/EDTA to extract ATP.

More recently, additional results have been published from Hastings and co-workers.[97-99] Seventy-six clinical isolates of *Enterobacteriacae* were tested against ampicillin, pipercillin, and gentamicin in a 4 h test with ATP extracted at room temperature. Correlation with MIC determinations was good.[97] They have also developed bioluminescence susceptibility assays for *Mycoplasma hominis* (after 24 h rather than the more usual 3 to 4d),[98] and for *Myco-bacteria* (after 5 d rather than several times that long).[99]

Investigations of Park et al. have also been recently extended to detecting methicillin-resistant *Staphylococci* in 4 h,[100] the antibiotic susceptibility of anaerobes in 5 h,[101] and the susceptibility of *Pseudomonas* to ciprofloxacin in 3 h.[102] Schifman et al. compared the results of the LAD Sensi-Quik® with those of the Abbott Avantage and the disk diffusion reference method.[103] Their results are presented in Table 4.

The LAD approach has been to try to develop a bioluminescence AST test that is practical for routine use in a clinical microbiology laboratory. We felt that use of boiling buffers or strong acids as ATP extractants was impractical in this setting, and that a fixed temperature and time of incubation were required.

E. THE LAD SENSI-QUIK®
1. Introduction

The LAD approach was to try to produce a product that would be accepted in the marketplace because of its speed, accuracy, sensitivity, and convenience. The initial goal was to develop a system that would deal with the most numerous clinical isolates. Later developments may include specialized media, different incubation times and temperatures, and other modifications needed to increase the test's utility.

The necessary studies to develop Sensi-Quik® were supported by $486,000 from the New Mexico Research and Development Institute, a branch of the New Mexico State Government, over an approximately 2-year period.

TABLE 4
Two-Hour ATP Production by *Staphylococcus aureus* in Different Broths

Broth	ATP Produced[a]		Ratio	
	Zero time	2 Hours	2 H/O time	Rank
Brucella	15,698	272,601	17	1
Mueller-Hinton	12,736	190,137	15	2
Columbia	14,519	199,038	14	3
Brain-Heart infusion	15,637	204,362	13	4
Eugonic	8,843	205,521	11	5
Thioglycolate	16,256	75,211	5	6
Typticase soy	44,131	205,339	5	6

[a] Expressed as Relative Light Units in counts per second.

We investigated the responses to 17 antibiotics of various combinations of 96 reference strains of bacteria (both Gram positive and Gram negative), obtained from the American Type Culture Collection (ATCC), and 374 clinical isolates (about half Gram negative) supplied by our collaborator, Ron B. Schifman, M.D., of the University of Arizona and the Veterans Administration Medical Center in Tucson.

Results of the bioluminescence AST were compared to the results obtained by the disk diffusion standard method in both Los Alamos and Tucson, and to the Abbott Avantage in Tucson. A current evaluation under way in Tucson adds comparison with the Vitek AMS, and an MIC determination.

Bioluminescence measurements were carried out using LAD reagents following a 3 h incubation of the bacterial suspensions with or without antibiotics at 35°C. The starting inoculum was about 1×10^5 cfu/ml, as determined by dilution of suspensions compared to a McFarland standard or prepared with the Prompt system. Three LAD luminometers were used, a model 535 (in both Los Alamos and Tucson), a model 735, and a model 633, both used in Los Alamos.

All three of the microtiter plate luminometers then available were evaluated. These were the Amerlite (from Amersham International, Little Chalfont, U.K.), the Dynatech (made by Torcon, Inc., Torcon, CA for Dynatech Corp.), and the Luminoskan, made by Labsystems Oy, Helsinki, Finland.

2. Medium Studies

Table 4 shows the results of a study on seven different media with the aim of maximizing ATP production by *S. aureus*. Similar studies were carried out on 11 other organisms. At least fivefold increases in ATP content (cell number) were seen in all media tested, while the best media gave an order of magnitude or better. When all the data were evaluated Columbia broth was the best of the seven media tested. For commercial purposes, however, we elected to use Mueller-Hinton broth supplemented with calcium and magnesium and diluted 1:1 with water. Mueller-Hinton is the medium recommended by NCCLS; the 1:1 dilution is thought to be necessary to produce cell lysis following the action of cell wall active antibiotics.

3. Ionic Strength Effects

We also investigated the effects of ionic strength on ATP production and antibiotic action. Table 5 shows the effect of decreasing tonicity (salt concentration) on ATP production for 13 ATCC strains.

TABLE 5
Effect of Decreasing Tonicity (Salt Concentration) on ATP
Production[a] Ratios (at 2 h vs. 0 h) in Columbia Broth

ATCC strains	5 g NaCl/l	2 g NaCl/l	1 g NaCl/l	0 g NaCl/l
K. pneumoniae	4.8X	3.1X	2.8X	2.7X
K. oxytoca	4.7	3.4	3.4	3.2
P. mirabilis	5.3	2.1	1.8	1.6
E. aerogenes	13	7.4	7.8	7.3
P. stuartii	5.2	2.1	1.8	1.6
E. cloacae	28	16	14	11
P. aeruginosa	1.8	1.4	1.4	1.3
S. aureus	14	7.2	6.6	6.8
S. saprophyticus	5.2	3.5	3.4	3.0
C. freundii	33	24	26	17
S. marcescens	37	21	19	14
E. coli	32	23	21	17
E. faecalis	28	21	20	16

[a]   Relative Light Units expressed as counts per second.

Reducing the added salt to zero reduced ATP production to approximately one half (of the 5 g/l level) for all organisms tested. Sluggish ATP producers (*Klebsiella, Providencia, Pseudomonas*, and *S. saprophyticus*), are also clearly identifiable.

We elected to test antibiotic effects on ATP production with 0 and 2 g NaCl/l because it had been demonstrated that a hypotonic medium enhances the action of antibiotics that act on the cell wall of bacteria. Sixty-one bacteria/antibiotic combinations were tested in Columbia broth containing 0 and 0.2% salt. The 2 g addition showed a modest superiority over broth with no NaCl, i.e., the antibiotics gave greater inhibition of ATP production (data not shown).

4. Incubation Times

Figure 1 shows that with *K. pneumoniae* treated with six different antibiotics, a clear distinction can be made between resistant and susceptible response in 90 min. Figure 2 presents the effects of the same antibiotics on *E. cloacae*. In this case, the resolution between susceptible, intermediate, and resistant is quite clear at 2 h. With ticarcillin, however, the organism appears resistant at 2 h, but intermediate at 3 h. In Table 6 the time course data is given for two antibiotics and six organisms. Five of the six bacterial strains tested showed a change in category response for one, or more often both, of the antibiotics. Similar results were obtained when an ATCC strain of *E. coli* was compared with four clinical isolates of *E. coli* and three antibiotics at 1, 2, and 3 h (data not shown). Our decision was to settle on a 3 h test.

5. Antibiotic Concentration Curves

Curves were constructed for ten ATCC strains growing at 35°C in the presence of three different concentrations of five antibiotics in cation-supplemented Mueller-Hinton broth. Representative data for gentamicin are shown in Figure 3. In each case, only 1 μg/ml of gentamicin reduces the Relative Light Output to less than 10% that of controls. Thus, all the organisms are susceptible to gentamicin at this concentration, a result confirmed using the Kirby-Bauer reference method.

6. Error Rates

An AST result is considered to be in error if it does not agree with one of the reference methods, i.e., Kirby-Bauer disk diffusion or broth dilution. Errors are of three kinds. A

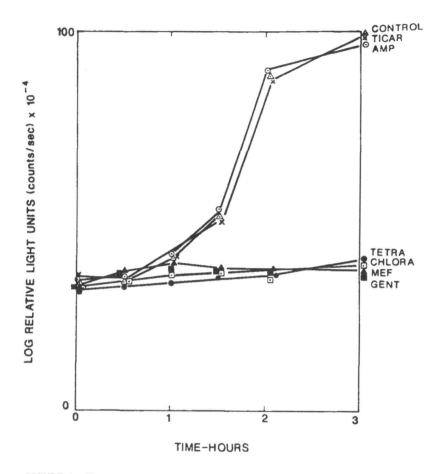

FIGURE 1. Time course study of the effects of six antibiotics on *Klebsiella pneumoniae*. The bacteria were suspended in Mueller-Hinton broth diluted 1:1 with water and supplemented with calcium and magnesium and incubated for 3 h at 35°C without antibiotics (control), or with ticarcillin, ampicillin, tetracycline, chloramphenicol, mefcillin, or gentamicin. Light output (in relative light units), was measured every 30 min.

very major error (VME) occurs when the test being evaluated says an organism is susceptible to an antibiotic while the reference method says it is resistant. This can mean that a patient is treated with a wholly ineffectual antibiotic, and is obviously the most serious outcome of an inaccurate test result.

A major error (ME) occurs when the test calls an organism resistant to an antibiotic where the reference method says it is susceptible. This means that a patient may be denied the use of an effective antibiotic because of an erroneous test result.

The final category is that of minor errors (MI) in which either the test or the reference method gives an intermediate result. The recommended standards are that the sums of VME and ME should be not more than 10%, or the even more stringent condition that the sum of all errors should not be more than 5%, and that VMEs should be 1.5% or less.[104,105]

Some antibiotic/organism combinations routinely generate a large number of errors. Among those that cause trouble in 4 to 8 h tests are[106]

Enterobacter tested against ampicillin or cephalothin
Enterococci vs. aminoglycosides or cephalothin
Serratia vs. tetracyclines
P. mirabilis vs. chloramphenicol

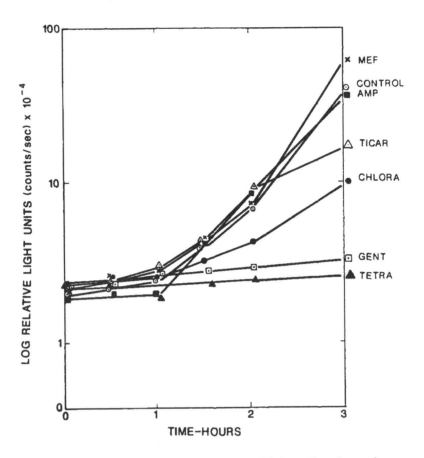

FIGURE 2.   Time course study of the effects of six antibiotics on *Enterobacter cloacae*.
The conditions were those of Figure 1.

TABLE 6
Comparison of Category (SIR)[a] Results at Different Incubation Times

	Ampicillin					Tetracycline				
	30	60	90 (min)	120	180	30	60	90 (min)	120	180
ATCC strains										
E. cloacae	R	—[b]	R	R	R	R	—	I	S	S
E. coli	R	R	R	S	S	—	I	I	I	S
K. pneumoniae	R	R	R	R	R	S	S	S	S	S
P. aeruginosa	R	R	R	R	R	—	R	R	S	R
S. marcescens	R	R	R	R	S	R	R	R	I	I
P. mirabilis	—	R	R	R	S	—	—	—	S	S

[a]   S = susceptible, I = intermediate, or R = resistant to the antibiotic.
[b]   ATP content decreased.

P. aeruginosa, *S. marcescens*, and *P. mirabilis* vs. aminoglycosides and trimethoprimsul-
famethoxazole
Citrobacter diversus, *K. pneumoniae*, *S. marcescens*, and *Enterobacter* vs. tetracycline

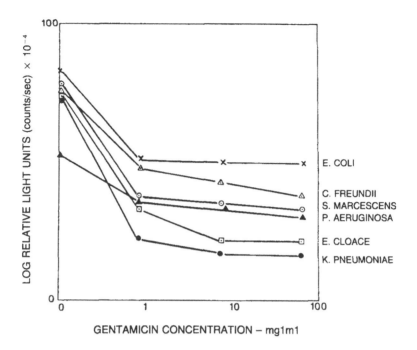

FIGURE 3. Effect of increasing gentamicin concentration on ATP production by six Gram negative bacteria. The conditions were those of Figure 1.

For 179 Gram negative rods tested against a small battery of our most troublesome antibiotics the errors were VME = 2.8 %, ME = 2.2, MI = 3.4 for a total error rate of 8.4%.

At the Veterans Administration Medical Center in Tucson, Dr. Schifman carried out a comparison of a preliminary form of what was to become Sensi-Quik© with the Abbott Avantage and the Kirby-Bauer method.[103] His results are shown in Table 7.

7. Microtiter Plate (MTP) Luminometer Evaluations

Microtiter plate luminometers offer a unique combination of small volumes (a microtiter plate well holds about 400 μl brim full), with a corresponding reduction in test costs, rapidity (96 wells can be read in a few minutes), and a high degree of automation for both specimen preparation and reading. The best of the microtiter plate luminometers possess considerable versatility (making them useful as research instruments as well as for routine analysis), and comparatively low cost.

The current disadvantages are that most of the instruments available are prototypes, and can be exasperating to use because of awkward programming and general unreliability. A more serious defect is that about an order of magnitude of sensitivity is lost compared to tube-based luminometers. This is to some extent inherent in the design since the area of light emission is necessarily limited to the surface area of a microtiter plate well.

There are only three MTP luminometer manufacturers at the moment, although this number will probably increase in the near future.

Amersham International (Amersham, U.K.) manufactures the Amerlite microtiter plate luminometer. This instrument is fully developed (it is not a prototype), and over 100 have been sold, largely for chemiluminescence immunoassays.

The instrument ran without a flaw from the moment we first plugged it in. Its most conspicuous problem from the standpoint of using it as a research instrument is its lack of versatility. All injections must be made on the laboratory bench before the microtiter plate is placed in the instrument, and there is at present no choice of programs to be run, or ways

TABLE 7
Discrepancies Compared to the
Reference Method(Kirby-Bauer)[a]

Very Major

	Avantage	LAD system
Ampicillin	3	0
Cephalothin	7	3
Gentamicin	3	1
Amikacin	0	0
Tetracycline	1	0
SXT[a]	1	0
Total	**16 (6.7%)**	**4 (1.7%)**

Major

	Avantage	LAD system
Ampicillin	1	0
Cephalothin	0	0
Gentamicin	0	0
Amikacin	2	0
Tetracycline	2	0
SXT[a]	0	1
Total	**5 (2.1%)**	**1 (0.4%)**

Very Major Errors Plus Major Errors

21 (8.8%)   5 (2.1%)

[a]   Number of specimens tested = 239.
[b]   Trimethoprim/sulfamethoxazole.

Note: Chi squared = 9.15, $p < 0.005$.

to adjust the instrument parameters (delay and integration times, for example) to suit the experimenter's convenience. It is, however, a well-designed, well-built, and handsome instrument that performs its deliberately limited functions extremely well.

Dynatech sells an MTP luminometer made for it by Torcon, of Torrance, CA. The Dynatech instrument originally had the ability to inject one reagent; the current model (the Microlite ML 1000, model 3) comes with three Cavro injectors controlled by the instrument's computer. The Dynatech is extremely versatile, offering a wide variety of operating modes, including plotting time course curves of the reaction, that are very welcome in the development laboratory.

The Dynatech instrument was the first that we used (in 1987), and it was the most prototypical of the instruments we tested. That is, it had many problems, mostly with temperature control of the electronics, and with the most difficult problem faced by the MTP luminometer designer, namely how to engineer a plate-moving system so powerful and precise that the injected reagent ends up in the right well and not alongside it, and does so for thousands and thousands of reading cycles.

We worked with three of these instruments over an 18 month period, and it was frequently traumatic. But each instrument was substantially better than the one before, so Torcon is effectively solving the problems. The Dynatech instrument should be evaluated by anyone considering the instrument's purchase of an MTP luminometer.

Luminoskan (Labsystems Oy, Helsinki, Finland) combines a well-designed, well-engineered, and handsome MTP luminometer with a battery of four peristaltic injectors. To

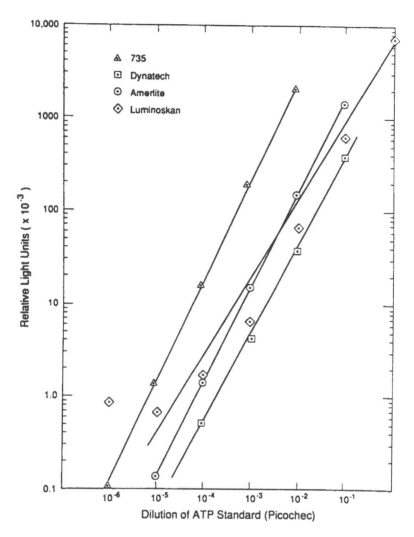

FIGURE 4. Standard curves (relative light units vs. dilutions of the ATP standard) for three microtiter plate luminometers and one tube luminometer (the LAD model 735).

our surprise, these injectors proved to be precise and reliable. The Luminoskan's problems were a very inflexible and inconvenient computer control system, and an electronic problem that eventually inactivated every system (three in all) that we tested. Labsystems has an impressive array of talent working to solve the Luminoskan's problems, and I have no doubt that an appealing and competitive instrument will result. It should be available early in 1990.

Standard curves were run on all three instruments and are presented in Figure 4. These data were obtained, for the most part, at different times and are not strictly comparable. They do, however, illustrate the properties of the three MTP luminometers compared to a standard tube luminometer (the LAD model 735). The loss of an order of magnitude in sensitivity in the MTP luminometers is clear from the data. This is better than what might be expected from the relative areas of a standard phototube window and the size of a microtiter plate well.

At the time these evaluations were made there was a software problem on the Dynatech luminometer that caused low counts to be printed incorrectly (Dr. Philip Stanley, personal communication).

In our AST studies, 100 Gram negative clinical isolates were run on the Luminoskan against ten antibiotics, and the results compared with the Kirby-Bauer reference method. Some of the data are shown in Table 8. Overall, the error rates were VME = 2.8%, ME = 4.0, MI = 8.6.

These data are close to meeting the criteria discussed earlier, except for the high rate of minor errors. As usual, the results can be manipulated by the proper choice of antibiotics and organisms (a caveat that applies to all AST data from any source). There were no errors in 100 tests with either imipenem or norfloxacin. Amikacin had no VMEs, and gentamicin had only one. At the other end of the spectrum, ampicillin had six VMEs, and tetracycline had four.

E. coli strains 1002, 1004, and 1005 (Table 8) produced 5 VMEs, 2MEs, and 2MIs, whereas *E. coli* strains 1008, 1019, and 1029; *S. marcescens* strain 3007, *K. pneumoniae* strains 3029 and 3040; *E. cloacae* strain 3033, *P. mirabilis* strain 3044; and *K. oxytoca* strain 3046 (data not shown), produced no errors when tested against any of the ten antibiotics.

Sensi-Quik™ shares with all other rapid AST tests a vulnerability to VMEs with beta lactam antibiotics and some organisms. These errors result because of the overnight induction of an enzyme (beta lactamase) that hydrolyzes the beta lactam group in the penicillins, and cephalosporins, thus rendering the antibiotic ineffective. Thus, Sensi-Quik™, like all other rapid AST tests, may show an organism to be susceptible to a beta lactam antibiotic when it is in fact resistant due to the induction of beta lactamase.

Imipenem is a beta lactam antibiotic but it is resistant to beta lactamase because of the configuration of the beta lactam ring side chain.

There are ways of dealing with the beta lactamase problem. The most common solution to the clinical problem is using the beta lactam antibiotics in combination with a beta lactam inhibitor. In addition, there are compounds that accelerate the induction of beta lactamase; these offer another line of attack on this problem in the laboratory environment.

8. Current Status

Sensi-Quik™ currently consists of boxes of 50 tubes of each of ten lyophilized antibiotics that can be combined to give a suitable panel at the convenience of the user. A reasonable panel for urine specimens might contain only four or fewer antibiotics depending upon the institution's practices for treating UTI.

More antibiotics will be added to the panel as soon as possible. We would hope to offer at least 24, and continue to increase the list as new antibiotics become available.

An ATP standard, blank tubes, lyophilized luciferin/luciferase, BRA, buffer, and tubes of cation-supplemented Mueller-Hinton broth with 5% lysed horse blood completes the kit.

In the standard procedure described in the package insert, an inoculum of approximately 10^5 cfu/ml of the organism to be tested is added to a broth tube, vortexed, and 25 µl aliquots of that suspension are added to the control tube and to the tubes containing the desired lyophilized antibiotics. An uninoculated broth tube is run as a negative control, and an ATP standard is run to insure that the system is working properly.

The tubes are incubated for 3 h at 35°C, then read in either the model 535 or the model 633 luminometer. The 633 has menus for both UTIscreen™ and Sensi-Quik™ that permit the technologist to enter a patient identification number, the infection site, the battery of antibiotics selected, and the origin of the specimen, i.e., whether it is from a plate, a direct positive urine, or a direct positive blood. The model 633 then prints out the results and indicates the category (SIR) for each antibiotic/organism combination.

Sensi-Quik™ is presently undergoing beta testing at the VA Medical Center in Tucson, and at Presbyterian Denver Hospital. Beta testing will be followed by clinical trials in the U.S. and Europe, and by submission to the FDA of a 510K (request for permission to market).

Other automated AST systems (the Vitek AMS, Abbott Avantage, Sensititre, Microscan)[107] are slower than Sensi-Quik™, and less flexible with regard to the antibiotic panel

TABLE 8
Luminoskan (L) Results vs. Kirby-Bauer (KB) for Gram Negative Rods

	Strain #	CHL L	CHL KB	IMI L	IMI KB	NOR L	NOR KB	SXT L	SXT KB	TET L	TET KB	AMP L	AMP KB	AMI L	AMI KB	CEF L	CEF KB	CEP L	CEP KB	GEN L	GEN KB
EC	1001	S	S	S	S	I	S	S	S	S	S	S	S	S	S	S	S	S	S	S	S
EC	1002	S	S	S	S	S	S	R	R	I	R	S	R	S	S	S	S	S	S	S	S
EC	1004	S	S	S	S	S	S	S	R	S	R	S	R	S	S	S	S	S	I	S	S
EC	1005	R	R	S	S	S	S	S	R	R	R	S	R	S	S	S	S	S	I	S	S
MM	2299	R	R	S	S	S	S	S	S	R	I	R	R	S	S	S	S	R	R	S	S
SM	2424	I	S	S	S	S	S	I	S	R	R	R	R	S	S	S	I	R	R	S	S
CF	2432	S	S	S	S	S	S	S	S	R	R	R	R	S	S	S	S	R	I	S	S
PM	3002	S	S	S	S	S	S	S	S	R	R	S	S	I	I	S	S	S	S	I	S
PS	3034	S	S	S	S	S	S	R	R	R	R	R	R	I	S	S	S	R	R	R	R
PV	3038	S	S	S	S	S	S	S	S	R	R	R	R	S	S	S	S	S	I	S	R
CD	3052	S	S	S	S	S	S	S	S	I	S	S	S	S	S	S	S	S	S	R	S

Note: Abbreviations are: CHL = chloramphenicol; IMI = imipenem: NOR = norfloxacin; SXT = trimethoprim/sulfamethoxazole; TET = tetracycline; AMP = ampicillin; AMI = amikacin; CEF = cefoxitin; CEP = cephalothin; GEN = gentamicin; EC = *E. coli*; MM = *M. morganii*; SM = *S. marcescens*; CF = *C. freundii*; PM = *P. mirabilis*; PS = *P. stuartii*; PV = *P. vulgaris*; CD = *C. diversus*.

available, but they are also cheaper. An antibiotic panel of one of these other automated instruments costs about $0.40, whereas Sensi-Quik™ will be about $1.80. We believe that Sensi-Quik™ will be used initially for life-threatening infections such as meningitis, septicemia, infective endocarditis, where rapid, effective treatment is essential to prevent death or serious complications. Later, it may become more widely adopted as physicians become accustomed to receiving results more rapidly than they are presently available.

IV. CONCLUDING REMARKS

What bioluminescence has principally to offer the clinical microbiologist is sensitivity. The practical translation of this in the laboratory is more rapid acquisition of test results. Realistically, we must ask ourselves, who cares?

The ultimate consumer of results from the clinical microbiology laboratory is the physician. A "rapid" result, to be potentially useful to the physician, should be available in patient care areas (preferably entered into or attached to the patient's chart) by approximately 4:00 p.m., and not later than 5:30 p.m., because, for effective utilization, the results must coincide with a physician-patient encounter.[108]

This means that for a "same day" result, i.e., a truly rapid test, on a specimen that arrives in the laboratory by 10:00 a.m., sample preparation, running of the test, and reporting of the results back to the appropriate nurse's station must be completed in approximately 6 h. This is easy for UTIscreen™ and the other urine screening methods. Most automated AST tests, however, that take 5 h to complete (i.e., Sensititre, Vitek AMS for about 70% of specimens) after the specimens are grown up on agar plates cannot be considered as same day tests. The Abbott Avantage and the Microscan Walkaway, that require 6 to 8 h are certainly not same day results by the criteria proposed by Jorgensen and Matsen.[108] The LAD direct Sensi-Quik™ test on positive urines would meet the same day requirement, as would the same direct test on positive blood specimens.

The LAD test as applied to specimens from plates is a second day test. Other automated AST tests that require specimens from plates report results on the third day after the specimen is received in the laboratory.

Another equally important problem is utilization of rapid results by the physician. In one study, only 52% of physicians were aware of the AST results on their patients 3 days after these results had been placed in the patient's charts.[109] This may seem to be a damning indictment of physicians, but, as the famous Gershwin song says, "It ain't necessarily so." Because of the slowness with which AST results are presently completed, overnight culture followed by an additional 18 to 24 h for Kirby-Bauer or broth dilution results, the physician begins treating the patient on the basis of clinical judgement. If the patient responds well to the antibiotic regime chosen, the physicians — pragmatists all — have no further interest in the susceptibility results. Only if treatment fails may the laboratory results become important.

A study based on retrospective chart review suggests that antibiotic therapy was inappropriate more than 40% of the time in hospitalized patients,[110] and that 32% of physicians started a different antibiotic upon receipt of a rapid preliminary AST result.[111] These data are consistent with the idea that physicians chose the right antibiotic first slightly more than half the time. A truly same day AST could greatly improve this situation, but only if communication between the laboratory and the physician is improved. The physician needs to know when he or she can expect a test result in order to plan effective treatment of the patient.[108]

A dozen infectious disease specialists voted 11 to 1 that they would prefer to have rapid, instrumented AST information rather than the more nearly definite overnight results.[112]

It may be appropriate to mention here that AST results, regardless of whether they are derived from rapid or classical tests, represent an "educated guess." They are generally run *in vitro* in defined aerobic media, on a single species at a low concentration (10^5 cfu/ml), and with pH control. In contrast, in an interabdominal infection from a perforated bowel, the bacterial level can be 10^{10} cfu/ml of mixed anaerobes and aerobes.[113] After an abscess forms, the bacteria are in an essentially anaerobic environment at an acidic pH. This is an extreme case, but predicting an outcome in the great variety of patient infections from any *in vitro* susceptibility test would, *a priori*, appear to be a hazardous procedure.

These limitations are well known. It is, however, true that treatment of infections usually fails when the physician uses an antibiotic to which the *in vitro* test says the infecting organism is resistant, and, conversely, treatment is usually successful if the physician uses an antibiotic to which the *in vitro* test says the infecting organism is susceptible.[114] Were this not so, antibiotic susceptibility testing would not make up such an imposing part of the workload of the clinical microbiology laboratory.

Although problems in its effective utilization remain, bioluminescence is poised to become an essential technique in clinical microbiology. Urinary specimens and AST comprise about 40% of the workload in clinical microbiology. Although bioluminescence will never be universally adopted for these purposes, it is having and will continue to have a substantial impact in the larger hospitals and reference laboratories because they are generally more open to new technology and more sophisticated about real costs than are many smaller groups. The ability to interface luminometer-associated computers with hospital and reference laboratory data management systems is becoming an increasingly appealing feature of the LAD systems to these large institutions.

For large laboratories (some of LAD's customers run 900 urine specimens a day), the microtiter plate luminometer with its capacity of running 500 specimens an hour or more is an appealing prospect, especially when Sensi-Quik™ can be run on the same instrument. Microtiter plate luminometer sensitivity is a problem with UTIscreen™; it is not with Sensi-Quik™.

Nucleic acid probes and monoclonal antibodies labeled with chemiluminescence or bioluminescence molecules can provide the sensitivity of radioisotopes without the associated disposal and shelf-life problems. Their role in microbial identification could play an important role in the clinical microbiology laboratory of the 1990s, and add another area where luminescence can make a definite contribution.

Less expensive reagents, more sensitive and convenient instruments, and more highly automated tests will help accelerate the penetration of bioluminescence into clinical microbiology.

ACKNOWLEDGMENTS

This manuscript has been critically read by J. G. M. Hastings, Ph.D., Department of Medical Microbiology, University of Sheffield Medical School, U.K.; Dr. Choong Park, Department of Pathology, Fairfax Hospital, Falls Church, VA; Ron B. Schifman, M. D., Director, Clinical Pathology, V.A. Medical Center and the University of Arizona, Tucson; and Philip E. Stanley, Ph.D., Consultant Scientist, Cambridge, U.K. They have labored mightily to reduce the errors of fact and interpretation in the original draft.

REFERENCES

1. **Weideman, E., II.** Ueber Fluorescenz und Phosphorescenz, I. Abhandlung, *Ann. Phys. Chem.*, 34, 446, 1888.
2. **Weideman, E. and Schmidt, G. C.,** 2. Ueber Luminescenz, *Ann. Phys. Chem.*, 54, 604, 1895. Weideman, E. und Schmidt, G. C., Ueber Luminescens von festen Körpern und festen Lösungen, *Ann. Phys. Chem.*, 56, 201, 1895.
3. **McCapra, F.,** The chemistry of bioluminescence, *Proc. R. Soc. London*, B, 215, 247, 1982.
4. **Dubois, R.,** Note sur la physiologie des Pyrophores, *C. R. Seances Soc. Biol.*, 2, 559, 1885.
5. **Leach, F. R.,** ATP determination with firefly luciferase, *J. Appl. Biochem.*, 3, 473, 1981.
6. **Gorus, F. and Schram, E.,** Applications of bio- and chemiluminescence in the clinical laboratory, *Clin. Chem.*, 25, 512, 1979.
7. **Whitehead, T. P., Kricka, L. J., Carter, T. J. N., and Thorpe, G. H. G.,** Analytical luminescence: its potential in the clinical laboratory, *Clin. Chem.*, 25, 1531, 1979.
8. **Harber, M. J.,** Applications of luminescence in clinical microbiology, in *Clinical and Biological Luminescence*, Kricka, L. J. and Carter, T. J. N., Eds., Marcel Dekker, New York, 1982, 189.
9. **Ulrich, P. G. and Wannlund, J. C.,** Bioluminescence in the microbiology laboratory, *Am. Clin. Prod. Rev.*, November, 50, 1984.
10. **Harber, M. J.,** Applications of luminescence in medical microbiology and hematology, in *Analytical Applications of Bioluminescence and Chemiluminescence*, Kricka, L. J., Thorpe, G. H. G., and Whitehead, T. P., Eds., Academic Press, London, 1984, 3.
11. **Hastings, J. G. M.,** Luminescence in clinical microbiology, in *Bioluminescence and Chemiluminescence, New Perspectives*, Schölmerich, J. et al., Eds., John Wiley & Sons, Chichester, England, 1987, 453.
12. **McElroy, W. D. and Deluca, M. A.,** Firefly and bacterial luminescence: basic science and applications, *J. Appl. Biochem.*, 5, 197, 1983.
13. **Wood, K. V., Lam, Y. A., Seliger, H. H., and McElroy, W. D.,** Complementary DNA coding click beetle luciferases can elicit bioluminescence of different colors, *Science*, 244, 700, 1989.
14. **Stannard, C. J. and Gibbs, P. A.,** Rapid microbiology: applications of bioluminescence in the food industry — a review, *J. Biolumin. Chemilumin.*, 1, 3, 1986.
15. **Stanley, P. E. and McCarthy, B. J.,** Reagents and instruments for use in the luciferase assay of ATP and luminescence: present needs and future possibilities, in *ATP Luminescence: Rapid Methods in Microbiology*, Society for Applied Microbiology, Tech. Ser., Vol. 26, Stanley, P. E., McCarthy, B. J., and Smither, R., Blackwell, Oxford, 1989, in press.
16. **Goodfield, C., Johnson, I. R., Massey, P. G., Stafford, D. A., Hampton, A. N., Campbell, A. K., and Newby, G. B.,** The use of highly sensitive luminometry for the selective detection of microbes, *op. cit.*
17. **Jago, P. H., Simpson, W. J., Denyer, S. P., Evans, Griffiths, A. W., Hammond, J. R. M., Ingram, T. P., Lacey, R. F., Macey, N. W., McCarthy, B. J., Salusbury, T. T., Senior, P. S., Sidorwocz, S., Smither, R., Stanfield, G., and Stanley, P. E.,** An evaluation of the performance of ten commercial luminometers, *J. Biolumin. Chemilumin.*, 3, 131, 73, 1989.
18. **Moyer, J. D. and Henderson, J. F.,** Nucleoside triphosphate specificity of firefly luciferase, *Anal. Biochem.*, 131, 187, 1983.
19. **McElroy, W. D.,** The energy source for bioluminescence in an isolated system, *Proc. Natl. Acad. Sci. U.S.A.*, 33, 342, 1947.
20. **McElroy, W. D. and Strehler, B. L.,** Factors influencing the response of the bioluminescence reaction to adenosine triphosphate, *Arch. Biochem. Biophys.*, 22, 420, 1949.
21. **Strehler, B. L. and Totter, J. R.,** Firefly luminescence in the study of energy transfer mechanisms. I. Substrate and enzyme determination, *Arch. Biochem. Biophys.*, 40, 28, 1952.
22. **Van Dyke, K., Stitzel, R., McClellan, T., and Szustkiewicz, C.,** An automated procedure for the sensitive ʳd specific determination of ATP, *Clin. Chem.*, 15, 3, 1969.
23. **ːpelle, E. W. and Levin, G. V.,** Use of the firefly bioluminescence reaction for rapid detection and counting of bacteria, *Biochem. Med.*, 2, 41, 1968.
24. **Picciolo, G. L., Kelbaugh, B. N., Chappelle, E. W., and Fleig, A. J.,** An automated luciferase assay of bacteria in urine, Goddard Space Flight Center, Report #X-641-71-163, 1971.
25. **Chappelle, E. W. and Picciolo, G. L.,** Adenosine triphosphate (ATP) as a possible indicator of extraterrestrial biology, NASA Technical Note D-7680, 1974.
26. **Vellend, H., Tuttle, S., Schrock, C. G., Deming, J. W., Barza, M., Weinstein, L., Picciolo, G. L., and Chappelle, E. W.,** Application of firefly luciferase assay for adenosine triphosphate (ATP) to antimicrobial drug sensitivity testing, NASA Technical Note D-8439, 1977.
27. **Chappelle, E. W., Picciolo, G. L., and Deming, J. W.,** Determination of bacterial content in fluids, in *Methods in Enzymology*, Vol. 57, Deluca, M., Ed., Academic Press, New York, 1978, 65.

28. **D'Eustachio, A. J. and Johnson, D. R.,** Adenosine triphosphate content of bacteria, *Fed. Proc.*, 1968, 761.

29. **D'Eustachio, A. J., Johnson, D. R., and Levin, G. V.,** Rapid assay of bacterial populations, *Bacteriol. Proc.*, 21, 13, 1968.

30. **Thore, A., Ansehn, S., Lundin, A., and Bergman, S.,** Detection of bacteriuria by luciferase assay of adenosine triphosphate, *J. Clin. Microbiol.*, 1, 1, 1975.

31. **Conn, R. B., Charache, P., and Chappelle, E. W.,** Limits of applicability of the firefly luminescence ATP assay for the detection of bacteria in clinical specimens, *Am. J. Clin. Pathol.*, 63, 493, 1975.

32. **Johnston, H. H., Mitchell, C. J., and Curtis, G. D. W.,** An automated test for the detection of significant bacteriuria, *Lancet,* August 21, 1976.

33. **Mackett, D., Kessock-Philip, S., Bascomb, S., and Easmon, C. S. P.,** Evaluation of the Lumac kit for the detection of bacteriuria by bioluminescence, *J. Clin. Pathol.*, 35, 107, 1982.

34. **Nichols, W. W., Curtis, G. D. W., and Johnston, H. H.,** Analysis of the disagreement between automated bioluminescence-based and culture methods for detecting significant bacteriuria, with proposals for standardizing evaluations of bacteriuria detection methods, *J. Clin. Microbiol.*, 15, 802, 1982.

35. **Curtis, G. D. W., Johnston, H. H., and Nichols, W. W.,** Bacterial ATP content of urines from bacteriuric and non-bacteriuric patients, in *Analytical Applications of Bioluminescence and Chemiluminescence,* Kricka, L. J., Thorpe, G. H. C., and Whitehead, T. P., Eds., Academic Press, London, 1984, 21.

36. **Drow, D. L., Baum, C. H., and Hirschfield, G.,** Comparison of the Lumac and Monolight systems for detection of bacteriuria by bioluminescence, *J. Clin. Microbiol.*, 20, 797, 1984.

37. **Nichols, W. W., Curtis, G. D. W., and Johnston, H. H.,** Detection of bacteriuria by bioluminescence: effect of pre-analysis centrifugation of specimens, *J. Appl. Bacteriol.*, 56, 247, 1984.

38. **Park, C. H., Hixon, D. L., McClintock, C. C., Ferguson, C. B., Lawless, C. C., Risheim, C. A., and Cook, C. B.,** Rapid detection of insignificant bacteriuria by concomitant use of Lumac system and Gram's stain, *Am. J. Clin. Pathol.*, 82, 593, 1984.

39. **Schifman, R. B., Weiden, M., Brooker, J., Chery, M., DelDuca, M., Norgard, K., Palen, C., Reis, N., Swanson, J., and White, J.,** Bacteriuria screening by direct bioluminescence assay of ATP, *J. Clin. Microbiol.*, 20, 644, 1984.

40. **Welch, W. D., Thompson, L. M., and Southern, P. M.,** Evaluation of two bioluminescence-measuring instruments, the Turner Design and Lumac systems, for the rapid screening of urine specimens, *J. Clin. Microbiol.*, 20, 1165, 1984.

41. **Bixler-Forell, E., Bertram, M. A., and Bruckner, D. A.,** Clinical evaluation of three rapid methods for the detection of significant bacteriuria, *J. Clin. Microbiol.*, 22, 62, 1985.

42. **Kolbeck, J. C., Padgett, R. A., Estevez, E. G., and Harrell, L. J.,** Bioluminescence screening for bacteriuria, *J. Clin. Microbiol.*, 21, 527, 1985.

43. **Males, B. M., Bartholomew, W. R., and Amsterdam, D.,** Leukocyte esterase-nitrate reductase and bioluminescence assays as urine screens, *J. Clin. Microbiol.*, 22, 531, 1985.

44. **Martin, E. T., Cote, T. A., Perry, L. K., and Martin, W. J.,** Clinical evaluation of the Lumac bioluminescence method for screening urine specimens, *J. Clin. Microbiol.*, 22, 19, 1985.

45. **Wu, T. C., Williams, E. C., Koo, S. Y., and MacLowry, J. D.,** Evaluation of three bacteriuria screening methods in a clinical research hospital, *J. Clin. Microbiol.*, 21, 796, 1985.

46. **Stamm, W. E., Hooton, T. M., Johnson, J. R., Johnson, C., Stapleton, A., Roberts, P. L., Moseley, S. L., and Fihn, S. D.,** Urinary tract infections: from pathogenesis to treatment, *J. Infect. Dis.*, 159, 400, 1989.

47. **Childs, S. J.,** Management of urinary tract infections, *Am. J. Med.*, 85, 14, 1988.

48. **Platt, R., Polk, B. F., Murdock, B., and Rosner, B.,** Mortality association with nosocomial urinary tract infection, *N. Engl. J. Med.*, 307, 637, 1982.

49. **Gregg, C. T., Leonard, P. A., and Eseke, J. R.,** Bioluminescence — the LAD UTIscreen kit, in *ATP Luminescence: Rapid Methods in Microbiology,* Society for Applied Bacteriology, Tech. Ser., Vol. 26, Stanley, P. E., McCarthy, B. J., and Smither, R., Eds., Blackwell, Oxford, 1989, 215.

50. **Pezzlo, M. T., Ige, V., Wollard, A. P., Petersen, E. M., and de la Maza, L. M.,** Rapid bioluminescence method for bacteriuria screening, *J. Clin. Microbiol.*, 27, 716, 1989.

51. **Baron, E. J., Tyburski, M. B., Almon, R., and Berman, M.,** Visual and clinical analysis of Bac-T-Screen urine screening results, *J. Clin. Microbiol.*, 26, 2382, 1988.

52. **Clarridge, J. E., Pezzlo, M. T., and Vosti, K. L.,** Cumitech 2A, *Laboratory Diagnosis of Urinary Tract Infections,* Weissfeld, A. S., Ed., American Society for Microbiology, Washington, D. C., 1987.

53. **Galen, R. S. and Gambino, S. R.,** *Beyond Normality,* John Wiley & Sons, New York, 1975.

54. **Joseph, J. M.,** Home test kits and mobile labs: strong reservations, *ASM News,* 55, 292, 1989.

55. **Bartlett, R. C.,** CAP workload recording, *Clin. Microbiol.*, 11, 68, 1989.

56. **Latham, R. H., Wond, E. S., Larson, A., Coyle, M., and Stamm, W. E.,** Laboratory diagnosis of urinary tract infection in ambulatory women, *JAMA,* 254, 3333, 1985.

57. **Mortensen, J. E. and Robinson, A.,** The use of Gram stain and leucocyte esterase/nitrite test strips for the detection of bacteriuria, *Lab. Med.,* 19, 364, 1988.

58. **Murray, P. R., Smith, T. B., and McKinney, T. C., Jr.,** Clinical evaluation of three urine screening tests, *J. Clin. Microbiol.,* 25, 467, 1987.

59. **Pezzlo, M.,** Detection of urinary tract infections by rapid methods, *Clin. Microbiol. Rev.,* 1, 268, 1988.

60. **Chaplin, M. F., Kozlov, I., and Barry, R. D.,** The preparation and properties of immobilized luciferase for use in the detection of bacteriuria, in *ATP Luminescence: Rapid Methods in Microbiology,* Society for Applied Bacteriology, Tech. Ser., Vol. 26, Stanley, P. E., McCarthy, B. J., and Smither, R., Eds., Blackwell, Oxford, 1989, 243.

61. **Filippova, N. Y., Dukhovich, A. F., and Ugarova, N. N.,** New approaches to the preparation and application of firefly luciferase, *J. Biolumin. Chemilumin.,* 4, 419, 1989.

62. **Ugarova, N. N., Brovka, L. Y., Lebedeva, O. V., Kutuzova, C. D., and Berezin, I. V.,** Immobilized polyenzymic bioluminescence systems for microassay, in *Bioluminescence and Chemiluminescence, New Perspectives,* Schölmerich, J., et al., Eds., John Wiley & Sons, Chichester, England, 1987, 583.

63. **Lundin, A.,** ATP assays in routine microbiology: from visions to realities in the 80's?, in *ATP Luminescence: Rapid Methods in Microbiology,* Society for Applied Bacteriology, Tech. Ser., Vol. 26, Stanley, P. E., McCarthy, B. J., and Smither, R., Eds., Blackwell, Oxford, 1989, 11.

64. **Worsfold, P. J. and Nabi, A.,** Bioluminescent and chemiluminescent assays using flow injection analysis, in *Bioluminescence and Chemiluminescence, New Perspectives,* Schölmerich, J., et al., Eds., John Wiley & Sons, Chichester, England, 1987, 543.

65. **McCapra, F. and Watmore, D.,** Synthesis and determination of some properties of firefly luciferin, *op. cit.*

66. **Simpson, W. J. and Hammond, J. R. M.,** Cold ATP extractants compatible with constant light signal firefly luciferase reagents, *op. cit.*

67. **Ugarova, N. N. and Dukovitch, A. F.,** Functions of lipids in firefly luciferase, in *Bioluminescence and Chemiluminescence, New Perspectives,* Scölmerich, J., et al., Eds., John Wiley & Sons, Chichester, England, 1987, 409.

68. **Schram, E. and Weyens-van Witzenberg, A.,** Improved ATP methodology for biomass assays, *J. Biolumin. Chemilumin.,* 4, 390, 1989.

69. **Leaback, D. H. and Haggart, R.,** The use of a CCD imaging luminometer in the quantitation of luminogenic immunoassays, *op. cit.,* 512.

70. **Fleming, A.,** On the antibacterial action of cultures of penicillium with specific reference to their use in isolation of *B. influenzae, Br. J. Exp. Pathol.,* 10, 226, 1929.

71. **Bauer, A. W., Kirby, W., Sherris, J. C., and Turck, M.,** Antibiotic susceptibility testing by a standardized single disk method, *Am. J. Clin. Pathol.,* 45, 493, 1966.

72. **Barry, A. and Thornsberry, C.,** Susceptibility tests: diffusion test procedures, in *Manual of Clinical Microbiology,* Lennette, E. H., Balows, A., and Hausler, W. J., Jr., Eds., American Society for Microbiology, Washington, D.C., 1985, 987.

73. National Committee for Clinical Laboratory Standards, (approved standard M2-A3), Performance standards for antimicrobial disk susceptibility tests, National Committee for Clinical Laboratory Standards, Villanova, PA.

74. **Ainsworth, K.,** Clinical impact of antibiotic susceptibility data, *Med. Lab. Observ.,* 19, 65, 1987.

75. **Doern, G., Scott, D., and Rashad, A.,** Clinical impact of rapid antimicrobial susceptibility testing of blood culture isolates, *Antimicrob. Agents Chemother.,* 21, 1023, 1982.

76. Albuquerque Journal, January 10, 1988, Associated Press.

77. **Ames, J. S.,** The Adenosine Triphosphate Assay for Determination of Bacteriuria and Antimicrobial Sensitivity, Unpublished thesis (M.S.), University of Wisconsin, Medical Microbiology Department, 1970.

78. **Hojer, H., Nilsson, L., Ansehn, S., and Thore, A.,** In vitro effect of doxycycline on levels of adenosine triphosphate in bacterial culture. Possible clinical applications, *Scand. J. Infect. Dis. Suppl.,* 9, 58, 1976.

79. **Thore, A., Nilsson, L., Hojer, H., Ansehn, S., and Brote, L.,** Effects of ampicillin on intracellular levels of adenosine triphosphate in bacterial cultures related to antibiotic susceptibility, *Acta Pathol. Microbiol. Scand. Sect. B,* 85, 161, 1977.

80. **Ansehn, S. and Nilsson, L.,** Effect of imidazole antifungals on *Candida albicans* demonstrated by bioluminescence assay of ATP, *op cit.,* p. 63.

81. **Hojer, H., Nilsson, L., Ansehn, S., and Thore, A.,** Possible application of luciferase assay of ATP to antibiotic susceptibility testing, in *Proc. Int. Symp. Anal. Appl. Biolumin. Chemilumin.,* Scram, E. and Stanley, P. E., Eds., State Printing and Publishing, Westlake Village, CA, 1979, 523.

82. **Fauchere, J. L., Simonet, M., Descamps, P., and Veron, M.,** Determination rapide de la concentration minimale inhibitrice d'un antibiotique par dosage de l'ATP bacterien, *Ann. Microbiol. (Inst. Pasteur),* 133, 293, 1982.

83. **McWalter, P. W.,** Determination of susceptibility of *Staphylococcus aureus* to methicillin by luciferin-luciferase assay of bacterial adenosine triphosphate, *J. Appl. Bacteriol.,* 56, 145, 1984.

84. **McWalter, P. W.,** Rapid susceptibility testing of *Staphylococcus aureus*, in *Analytical Applications of Bioluminescence and Chemiluminescence*, Kricka, L. J., Ed., Academic Press, New York, 1984, 17.

85. **Beckers, B. and Lang, H. R. M.,** Rapid determination of bacteriuria in bacteriological routine laboratory: comparison between bioluminescence method and culture technique, *Zentralbl. Bakteriol. Parasitenk. Infektionskr. Hyg. Abt. 1 Orig. Reihe A,* 254, 515, 1983.

86. **Beckers, B., Lang, H. R. M., and Beinke, A.,** Determination of intracellular ATP during growth of *Escherichia coli* in the presence of ampicillin, in *Bioluminescence and Chemiluminescence, New Perspectives,* Schölmerich, J., et al., Eds., John Wiley & Sons, New York, 1987, 67.

87. **Kouda, M., Ouchi, Y., Takasaki, Y., Matsuzaki, H., Maeda, T., and Nakaya, R.,** Bioluminescent assay as a potential method of rapid susceptibility testing, *Microbiol. Immunol.,* 309, 1985.

88. **Barton, A. P.,** A rapid bioluminescent method for the determination of methicillin-resistance in *Staphylococcus aureus* colonies, *J. Antimicrob. Chemother.,* 15, 61, 1985.

89. **Schifman, R. and Delduca, M.,** Abstract C-85, Rapid susceptibility testing of *Enterobacteriacae* by bioluminescence of ATP, *Abstr. Ann. Meet. Am. Soc. Microbiol.,* Las Vegas, NV, March, 1985, 314.

90. **Park, C. H.,** Abstract C-86, Rapid bioluminescence method (3h) for determining the susceptibility of *Candida* isolates to Amphotericin B, *op. cit.,* 314.

91. **Wheat, P. F., Oxley, K. M., Spencer, R. C., and Hastings, J. G. M.,** Rapid antibiotic susceptibility testing by ATP bioluminescence assayed with a new luminometer, in *Bioluminescence and Chemiluminescence, New Perspectives,* Schölmerich, J., et al., Eds., John Wiley & Sons, New York, 1986, 495.

92. **Hastings, J. G. M., Wheat, P. F., Oxley, K. M., and Spencer, R. C.,** Abstract No. 330, ATP bioluminescence testing using a novel luminometer, *Abstr. 27th Intersci. Conf. Antimicrob. Agents Chemother.,* New York, 1987, 153.

93. **Wheat, P. F., Oxley, K. M., Spencer, R. C., and Hastings, J. G. M.,** Abstract P-14, Antibiotic susceptibility testing using ATP bioluminescence assayed on the Amerlite analyzer, *Abstr. 5th Int. Symp. Rapid Methods Automation Microbiol. Immunol.,* Florence, 1987, 281.

94. **Franc, B.,** The bioluminescence technique as a method for determining the sensitivity of microorganisms to antibiotics, in *Bioluminescence and Chemiluminescence, New Perspectives,* Schölmerich, J., et al., Eds., John Wiley & Sons, New York, 1986, 507.

95. **Nilsson, L. and Ansehn, S.,** Bioluminescence assay for studies of effects of antimicrobial agents on bacteria and fungi, *op. cit.,* 491.

96. **Nilsson, L., Hoffner, S. E., and Ansehn, S.,** Bioluminescence assay for rapid susceptibility testing of *Mycobacterium tuberculosis, op. cit.,* 503.

97. **Wheat, P. F., Hastings, J. G. M., and Spencer, R. C.,** Rapid antibiotic susceptibility tests on *Enterobacteriacae* by ATP bioluminescence, *J. Med. Microbiol.,* 25, 95, 1988.

98. **Wheat, P. F., Limb, D. I., Spencer, R. C., and Hastings, J. G. M.,** Abstract #1213, Antimicrobial susceptibility testing of *Mycoplasma hominis* by ATP bioluminescence, *Abstr. 28th Intersci. Conf. Antimicrob. Agents Chemother.,* Los Angeles, 1988, 326.

99. **Hastings, J. G. M., Sparham, P., Spencer, R. C., and Wheat, P. F.,** Abstract #1214, Susceptibility testing of Mycobacteria (MB) by ATP bioluminescence, *op. cit.,* 326.

100. **Park, C. H., Hixon, D. L., McLaughlin, C. M., and Cook, J. F.,** Rapid determination (4h) of methicillin-resistant *Staphylococcus aureus* using a bioluminescence method, *J. Clin. Microbiol.,* 26, 1223, 1988.

101. **Hixon, D. L. and Park, C. H.,** Abstract #1212, Rapid (5h) anaerobic susceptibility testing of *Bacteroides fragilis* group to cefoxitin, clindamycin, and penicillin by a bioluminescence method, *Abstr. Am. Soc. Microbiol. Ann. Meet.,* Miami Beach, FL, 1988, 326.

102. **Park, C. H. and Hixon, D. L.,** Abstract #1212, Rapid (3h) susceptibility testing of *Pseudomonas* species to ciprofloxacin by a bioluminescence technique, *op. cit.,* 326.

103. **Schifman, R. B., Reid, L. L., Leonard, P. A., and Gregg, C. T.,** Abstract #1211, Rapid susceptibility testing of *Enterobacteriacae:* bioluminescence and Abbott Avantage method compared, *Abstr. 28th Intersci. Conf. Antimicrob. Agents Chemother.,* Los Angeles, 1988, 326.

104. **Thornsberry, C.,** Automated procedures for antimicrobial susceptibility tests, in *Manual of Clinical Microbiology,* 4th ed., Lennette, E. H., Balows, A., Hausler, W. J., Jr., and Shadomy, H. J., Eds., American Society for Microbiology, Washington, D.C., 1985, 1015.

105. **Sherris, J. C. and Ryan, K. J.,** Evaluation of automated and rapid methods, in *Rapid Methods and Automation in Microbiology,* Tilton, R. C., Ed., American Society for Microbiology, Washington, D.C., 1982, 1.

106. **Hansen, S. L. and Freedy, P. K.,** Concurrent comparability of automated systems and commercially prepared microdilution trays for susceptibility testing, *J. Clin. Microbiol.,* 17, 878, 1983.

107. **Jorgensen, J. H.,** Instrument systems which provide rapid (3- to 6-hr) antibiotic susceptibility results, in *Automation in Clinical Microbiology,* Jorgensen, J. H., Ed., CRC Press, Boca Raton, FL, 1987, 85.

108. **Jorgensen, J. H. and Matsen, J. M.,** Physician acceptance of rapid microbiology instrument test results, in *Automation in Clinical Microbiology,* Jorgensen, J. H., Ed., CRC Press, Boca Raton, FL, 1987, 209.

109. **Matsen, J.,** Rapid reporting of results — impact on patient, physician, and laboratory, in *Rapid Methods and Automation in Microbiology,* Tilton, R. C., American Society of Microbiology, Washington, D.C., 1982, 98.

110. **Maki, D. G. and Schuna, A. A.,** A study of antimicrobial misuse in a university hospital, *Am. J. Med. Sci.,* 275, 271, 1978.

111. **Matsen, J. M.,** Means to facilitate physician acceptance and use of rapid test results, *Diagn. Microbiol. Infect. Dis.,* 2, 735, 1985.

112. **Matsen, J. M.,** How rapid should rapid methods be? *Clin. Microbiol. Newslet.,* November, 1982.

113. **Bohnen, J. M., Matlow, A. G., Mustard, R. A., Christie, N. A., and Kavouris, B.,** Antibiotic efficacy in intraabdominal sepsis: a clinically relevant model, *Can. J. Microbiol.,* 34, 323, 1988.

114. **McCabe, W. R. and Treadwell, T. L.,** In vitro susceptibility tests: correlations between sensitivity testing and clinical outcome in infected patients, in *Antibiotics in Laboratory Medicine,* Lorian, V., Ed., Williams and Wilkins, Baltimore, 1986, 925.

Chapter 2

RAPID ESTIMATION OF MICROBIAL NUMBERS IN DAIRY PRODUCTS USING ATP TECHNOLOGY

M. W. Griffiths

TABLE OF CONTENTS

I.   Introduction 30

II.   Estimation of Raw Milk Quality Using ATP Technology 30
      A.   Native Milk ATP 31
           1.   Location and Concentration of Native Milk ATP 31
           2.   Factors Affecting Levels of Native Milk ATP 31
           3.   Removal of Native Milk ATP 34
      B.   Somatic Cell ATP 35
           1.   Levels of ATP in Somatic Cells 35
           2.   Removal of Somatic Cell ATP 36
      C.   Bacterial ATP 37
           1.   Extraction of ATP from Bacteria in Milk 37
      D.   Hydrolysis of Bacterial ATP 38
           1.   ATPases Added to the Reaction Mixture 38
           2.   ATPases Naturally Present in the Milk Sample 38
                a.   ATPases from Somatic Cells 39
                b.   ATPases from Bacterial Cells 39
      E.   Quenching Effects of Milk on Light Emission 39
      F.   Instrumentation and Reagents 40
      G.   Techniques for Enumerating Bacteria in Raw Milk 40
           1.   Methods Involving Extraction and Hydrolysis of
                Nonbacterial ATP 40
           2.   Methods Involving Concentration of Bacteria 41
                a.   Concentration by Filtration 41
                b.   Concentration by Other Methods 43

III. Estimation of Pasteurized Milk Quality Using ATP Technology 44
      A.   Methods to Detect Product Quality 44
      B.   Suitability of Plate Count as Comparison for Bacterial ATP
           Assay 45
      C.   Hygiene Monitoring of Processing Plant 46

IV.   Sterility Testing of UHT Products 47
      A.   Comparison of Sterility Test Methods 47
      B.   Detection of Milk-Degrading Enzymes by Bioluminescence 47
           1.   Proteases 48
           2.   Lipases 48

V.   Monitoring the Activity of Starter Cultures 48
      A.   Bioluminescence to Monitor Starter Activity 48
      B.   Novel Methods to Detect Bacteriophage 50

VI.   Detection of Bacteria in Other Dairy Products 50

VII.   Detection of Specific Groups of Bacteria 52
        A.   *Lux* Gene Technology 52
        B.   Other Selective Methods 53
                1.   Differential Growth 53
                2.   Differential Extraction 54

VIII.   Conclusions 55

Addendum 55

References 56

I. INTRODUCTION

As with other sectors of the food industry, milk processors have sought more rapid methods to assess raw material and product quality. Such methods are of particular interest to the dairy industry, as exceptionally rapid techniques which render results within a 10 min time frame are required to assess the quality of incoming raw materials. This will allow individual processors to obtain data on the bacterial load of tanker milk within the turnaround time of the transport. Current methods only allow a historical assessment of quality. The introduction of rapid bacterial counting procedures will also be of assistance in process control, thereby reducing wastage due to inadequate product quality and enabling more efficiently planned production. Warehousing costs may also be reduced because stock awaiting microbiological clearance could be released earlier.

Of all the novel microbial counting procedures introduced over the last 20 years or so, the one which arguably offers the best chance of an extremely rapid result is based on estimation of microbial ATP. The principle of the method for use to estimate biomass has been outlined in previous chapters and will not be described further in this review. An attempt will be made to highlight the possible applications of bioluminescence methodology to the dairy industry, indicating advantages and pitfalls.

II. ESTIMATION OF RAW MILK QUALITY USING ATP TECHNOLOGY

The determination of ATP levels in milk contaminated by bacteria was first described in 1970 by Sharpe et al.[1] These workers used butanol/octanol to extract the ATP in raw milk. Levels of ATP in milk containing 1.1×10^4 bacteria per g averaged 1.68×10^7 fg/g. Using cultures of *Staphylococcus aureus* and *Lactobacillus acidophilus*, Sharpe et al.[1] found the concentration of ATP per bacterial cell approximated 1 fg. It was concluded that the ratio of intrinsic ATP to bacterial ATP in the milk was 1500:1. When the bacterial count in this milk increased to 2.6×10^6/g, the ATP concentration rose to 3.43×10^7 fg/g and the assumed ratio of intrinsic ATP to bacterial ATP fell to 15:1. However, calculations suggest that this corresponds to an ATP concentration of 6.76 fg per cell. When known amounts of *L. acidophilus* cells were added to milk, the increase in ATP concentration was coincident with a bacterial cell ATP of 0.33 fg. These apparent differences in intracellular ATP levels may be due to the stage of growth of the cells at the time of the estimation. It is known that levels of ATP per cell in *Streptococcus faecalis* varied according to stage of growth[2] and changes in intracellular ATP according to the point reached in their growth cycle have also been described for a number of bacteria grown in milk.[3]

Despite these problems, bioluminescence has been used with varying degrees of success to determine bacterial numbers in milk (Table 1). However, the lower limit of detection in the majority of cases seems to be about 1×10^6 cfu/ml. This is not adequate for use in the U.K. where average bacterial counts of ex-farm bulk tank milk are approximately 2×10^4 cfu/ml. Also, the recently introduced Milk Designations[4] recommend a standard of 1×10^5 cfu/ml for raw milk quality. Thus, any rapid method for enumerating bacteria in milk must be able to detect at least 1×10^4 bacteria per ml. The lack of sensitivity of the techniques so far described in the literature can be ascribed to a number of factors listed in Table 2. The main limitation governing sensitivity of the ATP methodology appears to be the high levels of nonbacterial ATP in milk.[1,5] This is not only true of milk but also of most foodstuffs.[6] The nonbacterial ATP in milk is derived from two sources viz. (1) native ATP secreted during lactation and (2) ATP present in somatic cells shed by the mammary gland.

A. NATIVE MILK ATP

1. Location and Concentration of Native Milk ATP

Very little information has been published on the concentration of free ATP in milk. Kay and Marshall[7] reported the presence of up to 33 mg of adenine nucleotide per l in goats' milk. In a study of bovine milk, Richardson et al.[8] suggested that ATP was unlikely to exist free in milk serum because of the presence of phosphatases and ATPases.[9,10] ATP was present at similar concentrations in whole milks and corresponding skim milks, but was not detectable in skim milk ultrafiltrates.[8] The adenine nucleotide was also not present in colloidal phosphate-free milk from which the calcium phosphate-citrate (CPC)-caseinate micelles had been removed. Griffiths[3] observed that 75% of the ATP remained associated with the casein fraction after ultrafiltration. It was postulated that the ATP was sequestered in the CPC complex[8] which is intimately associated with the casein micelles of bovine milk.[11] Levels of ATP in the nine milks studied by Richardson et al.[8] varied between 0.13 and 0.31 μmol/ l, with an average of 0.23. Values of ATP in bovine milk of less than 1 μM were quoted by Zulak et al.[12] The background level of ATP present in typical milk samples with low bacterial and somatic cell counts was 100 μg/l[13] similar to that found by Richardson et al.[8]

2. Factors Affecting Levels of Native Milk ATP

The level of the indigenous ATP in milk is affected by a number of factors. The level of nonbacterial ATP varied with stage of lactation.[3,14] The concentrations were highest at the beginning of lactation and peaked at a level of 600 pg/ml 40 days after parturition.[3] Data from a larger number of animals with a greater frequency of sampling[14] showed that the peak occurred within 10 days of the onset of lactation and values of ATP between 10 and 17 mol/ml were found. In both studies, the ATP levels declined sharply after this initial peak and reached a basal value about 90 days from the start of lactation. Griffiths[3] suggested that these changes simply reflected the metabolic activity of the mammary gland, as they paralleled changes in milk yield observed with stage of lactation. ATP levels within the mammary gland have been shown to be highest in the lactating gland and concentrations fell toward the end of lactation.[15] The ATP:AMP ratio was highest in the mammary glands of cows giving high milk yields.[15] Unlike Griffiths,[3] who only assayed the indigenous ATP, Emanuelson et al.[14] assayed total nonbacterial ATP by extracting the somatic cell nucleotide with Triton® X-100.[16] Thus, the changes they observed could merely reflect variations in the somatic cell population which occurred during lactation.

TABLE 1

Studies on the Use of Bioluminescence to Enumerate Bacteria in Raw Milk: Conditions of Assay

Nonbacterial ATP extraction and hydrolysis	Bacterial ATP extractant	Time (min)	No. of samples	Sensitivity[a]	Regression coefficient (r)	Ref.
NRS + apyrase + EGTA	L-NRB	6	182	5×10^6	0.84	23
NRS + apyrase	L-NRB	45	60	7×10^6		13
NRS + apyrase[b]	L-NRB	10	30	5×10^6		13
Trypsin + Lubrol PX followed by filtration through 0.4 μ nuclepore filter	L-NRB	20	14	1×10^5	0.95	13
NRS + apyrase	L-NRB	45	39	1×10^6	0.74	59
NRS + EDTA + apyrase	L-NRB	5	500	1×10^6	0.85	22
NRS + EDTA + apyrase	L-NRB	15 + 45	48	6×10^5	0.93[c]	21, 86
			209	3×10^5	0.93[d]	21, 86
Somatic cell extractant + apyrase followed by filtration	Bacterial extractant	52	78	1×10^5	0.84	89
			85	1×10^5	0.90	89
NRS + EDTA + apyrase	L-NRB	45	15	1×10^6	0.82	28
ATP difference method[e]	L-NRB	5	145	5×10^5		61
NRS + EDTA + apyrase	L-NRB	5	75	5×10^6		61
NRS + EDTA + apyrase	L-NRB			1×10^5		19
None	n-Butanol, MOPS buffer, n-Octanol	3	>50	1×10^6	0.62[f]	87
NRS + apyrase	L-NRB	25	240	1×10^5	0.81	58
Preheated to 40°C lysis + filtration[g]	Extractant	3	294	1×10^4	0.92	82
NRS + apyrase	L-NRB	45	33[h]		0.17	60
			50[h]		0.34	60
			45[h]		0.31	60
			14[h]		0.46	60
NRS for filtration + apyrase	NRM	7	285	2×10^4	—	64
Somex A + filtration	Bactex	8	240	5×10^3	0.64	170

a   Lowest number reliably detected.
b   Apyrase at 10 × concentration of original assay (Bossuyt 1981, 1982).
c   Tanker milk.
d   Farm milk samples.
e   Half of milk tested by NRS-EDTA-somase and L-NRB and other half with water substituted for L-NRB.
f   Determined for all samples.
g   In BactoFoss machine.
h   Range of counts 1×10^3 to 2.5×10^6; 5×10^2 to 3×10^5; 0 to 1×10^6 and 1×10^3 to 2.5×10^5.

TABLE 2
Factors Affecting the Sensitivity of the Bacterial ATP in Milk

Failure to extract and hydrolyze all free and somatic cell ATP
Failure to extract all bacterial ATP
Presence of somatic cell and possibly bacterial ATPases which hydro-
  lyze extracted bacterial ATP
Residual ATPase used for hydrolysis of nonbacterial ATP may decrease
  levels of bacterial ATP
Quenching effect of milk on luciferase reaction
Sensitivity of instrumentation
Purity of reagents, especially luciferase and luciferin

Experiments with goats' milk have shown that the ATP concentration in freshly secreted milk was three times greater than in milk which had been left in the udder overnight.[12] It was also suggested by these workers that caprine milk was capable of synthesizing ATP. The ATP content of goats' milk (12.4 to 37.6 μM) was significantly higher than that of bovine milk, determined by the same authors. It is uncertain whether bovine milk has the capability of synthesizing adenosine nucleotides.

Zulak et al.[12] found that 70% of the ATP in goats' milk which had accumulated in the udder overnight was present in the skim milk fraction. The nonfat component of freshly secreted milk contained only 33% of the nucleotide. It was suggested that the ATPase of the milk fat globule may degrade the fat-associated ATP with time. This would result in a lowering of the ATP content of the fat phase with a concomitant increase of the proportion observed in the skim milk. Alternatively, the ATP in the milk fat globule may be located in "cytoplasmic crescents".[17] These may preferentially degrade the high amounts of ATP they contain or the crescents could disintegrate with time in the udder releasing ATP into the aqueous phase of the milk. In freshly secreted milk more of the milk fat globule crescents remain intact and retain ATP. Less ATP would, therefore, be found in the skim milk.

3. Removal of Native Milk ATP

Firefly luciferase is an extremely hydrophobic protein[18] and, thus, may be able to penetrate the casein micelle where it would react with the CPC-bound ATP, resulting in light emission. Therefore, in order to assay the bacterial ATP, the native milk ATP must be extracted and hydrolyzed. It has been suggested that the poor correlation between ATP concentration and bacterial numbers in milk samples with counts less than 1×10^5 cfu/ml was due to the high levels of ATP associated with casein micelles.[19] The micellar ATP appears to be inaccessible to the nucleotide-hydrolyzing enzyme apyrase.[20] Ways must be sought to open up the molecular structure surrounding the micellar ATP to allow enzymatic hydrolysis. The bound ATP is intimately linked with the micellar calcium-phosphate-citrate, so sequestrants have been used to aid release of the nucleotide.[21-23] The casein micelles are dissociated by chelating agents,[24] with the result that the ATP bound to the calcium phosphate becomes more accessible to the enzyme apyrase. Incorporation of the chelating agent, ethylenediaminetetraacetate (EDTA), into the extraction mixture used by Theron et al.[25] to remove nonbacterial ATP from milk caused a decrease in background nucleotide levels. Similar results were achieved when a number of sequestrants were compared for releasing the micellar ATP.[23] These included ethylene glycol-bis(β-aminoethyl ether)N,N,N′,N′-te-traacetic acid (EGTA) and citrate as well as EDTA. Background counts appeared lower in the presence of citrate. This agreed with earlier findings and may have been due to inhibition of the luciferase enzyme.[19] However, citrate could compete with ATP for the colloidal phosphate of the casein micelle. It is known that citrate readily co-precipitates with calcium phosphate from alkaline/neutralized solutions to form amorphous gels.[26,27] Light emission tended to be highest in the presence of EDTA, especially in high bacterial count milks, although there was no overall improvement in the sensitivity of bacterial detection.[23] The concentration of EDTA present in the reaction mixture, however, was crucial,[28] and seemed to be optimal at 10 mM. Variation in results obtained using different batches of reagent was probably due to small differences in EDTA concentration. These concentration effects were not so apparent when EGTA replaced EDTA in the assay system, and this permitted higher concentrations of the chelator to be used.[23,29]

In order to make milk filterable through bacteria-retaining membranes, treatment with a protease and detergent is needed.[30,31] The protease partially digests the casein component of the milk and the detergent helps disperse the fat. A protease pretreatment, therefore, may be of benefit in liberating micellar-bound ATP. An improvement in destruction of micellar ATP was achieved by digestion with a proteolytic enzyme, pronase, especially when used in conjunction with sonication.[19] Care must be taken in the selection of the protease because some may denature the luciferase and also the apyrase needed to hydrolyze the nonbacterial ATP. The luciferase is particularly susceptible to degradation by trypsin,[29] but the apyrase was resistant to a number of proteases including trypsin, chymotrypsin, subtilisin, papain, and proteinase K.[32] Additionally, there may be an effect of such pretreatments on the bacterial membrane causing release of intracellular ATP that would in turn be hydrolyzed along with the nonbacterial ATP.

ATP sequestered on proteins and other macromolecules has been released by bringing the pH of the sample to 4.25 using malic acid and may be removed by apyrase, which remains marginally active at this pH.[33] However, problems may be encountered with milk samples because casein is precipitated at pH 4.6. A large proportion of the bacteria present (as much as 90%) remain closely associated with the precipitated casein and will affect levels of bacterial ATP subsequently assayed.

B. SOMATIC CELL ATP
1. Levels of ATP in Somatic Cells

All fresh raw milks contain cells derived from the mammary gland of the cow. These somatic cells, like all living cells, contain ATP. The levels of somatic cells in milk vary according to the disease status of the animal and rise to very high concentrations in milks from mastitic cows.[34] Somatic cell populations in excess of 1×10^7 may be observed in milks from cows with subclinical mastitis, although average values for ex-farm bulk tank milks in the U.K. are about 3×10^5.[35] As might be expected, the disease status of a group of animals is reflected by the ATP concentration found in their milk. Emanuelson et al.[14] quoted average figures of 3.52 mol ATP per ml for milk from a group of cows, but this value fell to 1.74 mol ATP per ml when only milk from healthy animals free from mastitic infection was analyzed. It has been suggested by a number of workers that ATP can serve as an index of mastitic infection.[14,16,36,37] A novel approach to the detection of mastitis using bioluminescence was proposed by Malkamaki et al.[38] These workers followed the increase in bacterial biomass in whole milk by monitoring ATP. Bacteria showed an enhanced replication rate in mastitic milks, and the increase in bacterial ATP correlated well with inflammatory markers for mastitis.

The ATP content of somatic cells is far in excess of that found in bacterial cells. Best estimates suggest that there are 10 to 100 times more ATP molecules in somatic cells compared to bacteria.[3,39] However, some workers have proposed that levels of adenine nucleotide in somatic cells may be 400 times those of bacterial cells[32] (Table 3). There is a wide variation in the ATP concentration found in somatic cells which may be due to methods used for nucleotide extraction and age of milk when tested. Olsson et al.[16] observed that significant amounts of cellular ATP were lost within 60 min from milk containing a high concentration of somatic cells, even when the milk was stored under refrigeration. The

TABLE 3
Concentration of ATP in
Somatic Cells

Approximate ratio of somatic: bacterial ATP/cell in milk	Ref.
26 : 1	3
600 : 1[a,b]	16
500 : 1	32
80 : 1[a,c]	36
10 to 100 : 1	39
500 : 1	58

[a]   Assuming an average of 1 fg ATP per bacterial cell.
[b]   Calculated from the data of Olsson et al.
[c]   From regression equation cited by Bossuyt.

effect was more pronounced at higher temperatures, with the result that there was an almost complete loss of ATP within 2 h at 50°C. This instability has also been noted by other researchers.[36,40] The ATP in milk containing low somatic cell populations was much more stable and presumably reflected the greater proportion of micellar-bound ATP found in these milks.[16] The fall in ATP levels was not necessarily correlated with changes in somatic cell count, as nucleotide levels were shown to decline more rapidly than cell numbers.[36]

Somatic cell numbers in milk are affected by several factors other than the disease status of the animal. Cell numbers are dependent on the stage of lactation and are highest in the period 1 to 10 days after parturition.[14,34,41-43] The somatic cell population then falls and remains constant until late in lactation (about 150 days after parturition), when numbers begin to increase again. Even in healthy animals free from mastitic infection variations in cell count between 7×10^4 and 7×10^5 can be observed during lactation.[14] This is reflected in changes in ATP concentration in milk at different stages of lactation,[3,14] though these may be partly due to changes in the micellar-bound ATP in milk.[3] However, the fluctuations in ATP concentration during lactation observed by Emanuelson et al.[14] followed exactly the changes in somatic cell count. ATP levels could vary by more than tenfold at different stages of lactation.[3,14]

As well as changes in somatic cell counts observed in milk during stage of lactation, there is also a tendency for cell counts to increase with increasing lactation number.[34,42-45] These variations were not so apparent in the study by Emanuelson et al.[14] Indeed, these workers observed that ATP concentrations were highest in milks from cows in their first lactation. The ATP levels in first-lactation animals did not rise later in the lactation period as they did in milks from cows undergoing subsequent lactations.

2. Removal of Somatic Cell ATP

As with the native milk ATP, the adenine nucleotide present in somatic cells must be extracted and hydrolyzed before bacterial ATP can be assayed. Lysis of somatic cells by detergents has been described,[46] and Bossuyt[36] used a commercial detergent system (NRS) which was claimed to be specific for extraction of somatic cell ATP. This treatment may not be completely effective in removing all the somatic cell nucleotide.[25] As much as 4% of the nonbacterial ATP remained after use of NRS-EDTA extractant systems, although it was difficult to assess the relative proportion of micellar-bound ATP contributing to this unextractable pool. Also, the NRS extractant may not be completely specific for somatic cells.[25] The ATP concentration per cell of the bacterium, *Enterobacter cloacae* was reduced by 65% when skim-milk-grown cultures of this organism were treated with NRS-EDTA at 30°C. This decrease may have been caused by release of some bacterial ATP, or interference with the nucleotide turnover mechanism of the intact bacterial cell. Theron et al.[25] concluded that this was due to a physiological change brought about by these reagents rather than a loss in viability.

Other nonionic detergents have been used to extract somatic cell ATP.[13,14,32] Olsson et al.[16] used a somatic cell extractant consisting of Triton® X-100, Tris buffer, and EDTA. They claimed that this was more effective than NRS, probably because of the instability of ATP in NRS and not because of greater extraction efficiency. However, other work has suggested that NRS gave a more rapid and complete release of mammalian cell ATP than Triton® X-100.[47] Bacteria did not increase ATP levels after Triton®/Tris/EDTA extraction which was claimed to be the result of the rigidity of the bacterial cell wall preventing lysis.[48] Unfortunately, Triton® X-100, as well as other detergents of the Triton® X series, inhibit luciferase.[32] This inhibition can be alleviated by protecting agents. Other nonionic detergents such as Lubrol-PX and Triton N-101 do not inhibit the luciferase reaction. These other nonionic detergents may also have an effect on bacterial cells, for example, Triton® X-100 at a concentration of 0.1% lysed cells of *Streptococcus*[33] and other bacteria, including *Pseudomonas aeruginosa*.[50] About 46% of the ATP content of *P. aeruginosa* was released by Triton® X-100. Triton® X-100 also inhibited subsequent growth of certain bacteria.[33]

A more complete degradation of nonbacterial ATP has been indicated after incubation at 50°C in the presence of Triton® X-100 and apyrase.[47,49] Heat treatments of this kind may lead to leakage of bacterial ATP.[5]

Naturally occurring glycoside detergents, saponins, can also be used to lyse somatic cells with little or no consequent effect on bacterial cells.[50,51] The preparation of stable cellular homogenates containing ATP by treatment with guanidinium chloride has been described,[52] but this system is probably of little use with milk.

C. BACTERIAL ATP

In order to be able to use bioluminescence for estimation of bacterial numbers it must be assumed that the ATP concentration per cell remains constant at a known value under defined conditions. This assumption is not strictly valid, as intracellular ATP levels in bacteria can be affected by a number of factors, including stage of growth cycle and environmental factors, among others.[33,53] Theron et al.[54] showed that the ATP content of mesophilic bacteria was considerably lower at refrigeration temperatures (4°C) than at 30°C. In the case of a psychrotrophic strain of *P. fluorescens* the ATP content was similar at both temperatures. Nevertheless, the ATP assay has been used successfully by numerous workers to determine bacterial concentrations in fluid samples.

1. Extraction of ATP from Bacteria in Milk

Stanley[53] outlined the requirements for an ideal bacterial extractant. However, in no case have they fully been met. When adenine nucleotides from a number of bacteria were extracted by ten different methods, it was concluded that extraction with trichloroacetic acid most closely reflected actual levels of the nucleotides in intact bacterial cells.[55] Difficulties are encountered when this extractant is used in milk. Proprietary extractants or "releasing agents" are available which require no postextraction processing.[56,57] These bacterial-releasing agents are generally cationic detergents.[32] By far the most widely used extractant for bacterial ATP assay in milk is L-NRB,[13,21-23,58-61] a mixture of ionic surfactants. Nucleotide release by this extractant is quantitative up to cell concentrations of 1×10^7 cfu/ml. For young cells it was concluded that NRB was a better extractant than TCA.[62] Although extraction of ATP is rapid with detergent systems (about 10 s), they often exert an inhibitory effect on luciferase. Also, ATP is unstable when dissolved in many of these detergents. Recently, protective agents have been described which prevent the deleterious action of cationic extractants on luciferase.[32,63] One such extraction system is commercially available (NRM; Lumit QM) and has been used to extract ATP from bacterial cells removed from milk by filtration.[64]

D. HYDROLYSIS OF BACTERIAL ATP

A source of error when estimating bacterial ATP by bioluminescence may be the loss of ATP due to enzymic hydrolysis. ATP-hydrolyzing enzymes are added to the reaction mixture to remove nonbacterial ATP and may also be present in somatic and bacterial cells to be released on lysis.

1. ATPases Added to the Reaction Mixture

Apyrase (an ATPase) was found to hydrolyze ATP in the intact cell either by destroying surface-associated ATP or by penetrating the cell.[25] Evidence that the treatment used to extract and hydrolyze nonbacterial ATP could affect bacterial ATP levels was obtained by Thore et al.[48] These workers reported that two bacterial species lost 50% of their intracellular ATP when treated with Triton® X-100 and apyrase. It was not made clear which component contributed to the fall in ATP levels. Similar results were achieved when NRS and apyrase were used to deplete nonbacterial ATP.[61] The reduction in intracellular bacterial ATP levels was strain dependent. In another study,[25] treatment of an *Enterobacter cloacae* culture in skim milk with apyrase (about 2 units per ml of culture) for 15 min at 24°C resulted in a 30% reduction of intracellular ATP compared with that found in culture where water had been added instead of enzyme. Under normal conditions of assay, 0.5 units per ml of apyrase are added to milk to hydrolyze nonbacterial ATP, and the milk sample is incubated for 5 min at room temperature.[22] Thus, using the above figures, a decrease of only about 2.5% in bacterial ATP may be expected under these conditions. The conclusion that the levels of apyrase added to milk to hydrolyze nonbacterial ATP does not greatly affect bacterial ATP levels is supported by the results of Webster et al.[13] They used apyrase levels tenfold greater than those previously quoted.[21,22] Under conditions that destroyed somatic cell ATP, there was no loss of bacterial ATP from either freshly grown cells of *Escherichia coli* or of an unidentified psychrotroph.

There is also the possibility that apyrase may hydrolyze the ATP on extraction from bacterial cells during the short time (usually 30 s) before luciferase addition. Apyrase is rapidly denatured by heating at 95°C.[25,65] Heat treatment of the reaction mixture at 95°C for 15 s prior to extraction of bacterial ATP with NRB did not have an appreciable effect on light output, suggesting little or no hydrolysis of the extracted ATP.[29] An alternative approach to ensure the absence of bacterial ATP degradation by apyrase is to use a noncompetitive inhibitor of the enzyme such as EDTA[55] or vanadate during the bacterial extraction step. Indeed, L-NRB itself partially inactivates apyrase,[66,67] as do most commercial extractants.[68] Other extractants such as dimethyl sulfoxide inhibit apyrase[69] but the sample has to be diluted about 60-fold to limit its effect on luciferase with a concomitant loss in sensitivity.

A further source of error may be the nonhydrolysis of extracted somatic cell ATP which may become bound to membranes and escape destruction by the apyrase enzyme.[47]

2. ATPases Naturally Present in the Milk Sample

ATP-hydrolyzing enzymes are present in somatic and bacterial cells. There is, therefore, potential for these enzymes to degrade bacterial ATP prior to assay.

a. ATPases from Somatic Cells

Botha et al.[66] showed that there was a curvilinear relation between somatic cell count and ATPase activity in milk. Detectable increases in ATPase levels coincided with somatic cell counts of approximately 3×10^5 cells per ml. Other workers have shown similar increases in ATPase activity at elevated cell counts.[70,71] The nonlinear nature of the regression line may, in part, be explained by changes in the proportion of types of somatic cell with count. Polymorphonucleocytes constitute a larger proportion of somatic cells in high cell count milks[72] and it is possible that enzyme activity is greater in these cells.[66] At somatic cell levels of 2×10^6/ml there was a 50% reduction in original ATP levels within 2 min of addition.[66] As the average cell count for milk from all marketing boards in the U.K. was below 3.6×10^5/ml in 1988,[73] the contribution of ATPases to the assay system from this source was probably insignificant.

Botha et al.[58] found that the standard error of estimate for assay of ATP in heated (110°C for 10 min) milk was lower than for raw milk. They proposed that raw milk contained a factor that was eliminated by heating. This factor was claimed to be very active ATPases from somatic cells which were denatured during the heat treatment.[74] There could equally have been an improvement in the precision of the method due to removal of somatic cell ATP which the heat treatment would have inevitably reduced.

The extractant most commonly used for extraction of ATP from somatic cells is NRS. This extractant quantitatively releases small molecules from cells while larger molecules, including the ATP-hydrolyzing enzymes, are retained. Thus, extracted nucleotide remains unaffected by these enzymes.

b. ATPases from Bacterial Cells

There was a linear relation between ATPase activity and bacterial count when a strain of *P. fluorescens* was grown in sterile milk at 7°C.[66] However, enzyme activity could not be detected until the bacteria had reached late exponential or early stationary phase of growth. At this time the count approached 1×10^8 cfu/ml. Similar results were obtained with *E. coli* when adenine nucleotides in the medium were only destroyed during late stationary growth phase.[75] Because of the high bacterial numbers required in milk before appreciable levels of ATPase are detected, it would seem unlikely that these enzymes affect the bacterial ATP assay.

E. QUENCHING EFFECTS OF MILK ON LIGHT EMISSION

Logarithmic plots of the light output by luciferase in various fluids containing ATP at concentrations between 6.25 and 200 pg yielded parallel linear responses.[76] The light output was dependent upon the solvent system used. The sensitivity of the assay was greatest when ATP was dissolved in water and lowest for ATP in skim milk. Similar results were obtained when ATP was present in water, raw, and heat-treated milk.[58] The lack of sensitivity in milk could be ascribed to chemical and optical interference or "quenching"[55,77-80] that was absent in distilled water.[36] Theron et al.[76] calculated that because of this quenching effect the least number of bacteria that could be detected in milk using their assay system was 1.56×10^5 bacteria per ml. Optimization of the milk bacteria assay originally described by Bossuyt[21,22] to minimize quenching of the light production by milk components and achieve greatest sensitivity only resulted in a 30% improvement.[13] The minimum number of bacteria detectable by this modified procedure was at best about 5×10^6.

The quenching effect can be readily measured and a correction applied by an internal standardization procedure. This is done by performing a second analysis on the sample to which has been added a known amount of ATP.[81] The increase in light emission registered in the second analysis is due to the added ATP and the percentage of this measured is directly related to the amount of quenching. This technique has not been widely adopted in methods described for the routine estimation of bacterial numbers in milk. It has been reported that correlations between bacterial count and light output in milk were lower if internal standardization was applied.[58] The authors claimed this was due to hydrolysis of the added ATP by ATPases of somatic cells.

F. INSTRUMENTATION AND REAGENTS

As with all new technologies, the sensitivity of instrumentation and purity of reagents for bioluminescence measurements are continually improving. Instruments offering advantages in terms of automation or portability are readily available and machines specifically designed for the food industry are appearing on the market.[82] Some desirable properties of luminometers for use in the food quality control laboratory were presented by LaRocco et al.[68] The performance of some commercial luminometers was recently evaluated[83] and differences in sensitivity established. However, performance of bacterial ATP assays in milk by a modification of Bossuyt's method[22] using instruments with a tenfold difference in sensitivity did not significantly improve results.[23] Jago et al.[83] concluded that the detection of low numbers of microorganisms can be enhanced significantly by carrying out several analyses on the same sample. This will be greatly facilitated by the availability of automated instruments.

Webster et al.[84] pointed out that most commercial firefly luciferase preparations are not saturated with luciferin and supplementation with this co-factor resulted in a 1.8-fold increase in light production. This addition seemed to have little effect on the sensitivity of the ATP assay in milk.[13] Griffiths and Phillips[23] compared two luciferase preparations and found that, although there were differences in light output between the enzymes, there was no difference in the sensitivity of the milk bacteria assay. A previous study[76] compared luciferase preparations with different sensitivities[84] at similar ATP concentrations. The results suggested that greater sensitivity of the luciferase-luciferin preparation improved the precision of the ATP assay in milk but would not necessarily enable lower numbers of bacteria to be detected.

Other methods of improving bioluminescence techniques including immobilization of reagents for use as a ''dipstick'' test are being studied.

Although improvements in instrumentation and reagents have made an impact by facilitating sample handling and increasing the precision of the assay procedure, before the full impact of the greater sensitivity of luminometers and enzyme preparations can be felt, better methods for removing background ATP from milk samples must be obtained.

Technical and biochemical parameters for ATP assay have been thoroughly reviewed by Lundin.[85]

G. TECHNIQUES FOR ENUMERATING BACTERIA IN RAW MILK

1. Methods Involving Extraction and Hydrolysis of Nonbacterial ATP

Several workers have studied bioluminescence for determining levels of bacteria in milk (Table 1). In most cases, the protocol follows closely that originally described by Bossuyt.[21] This involves extracting and hydrolyzing the nonbacterial ATP present in milk with a commercially available nonionic detergent of undisclosed composition (NRS) containing a sequestrant (usually EDTA) and the enzyme apyrase. The bacterial ATP is then extracted with a second detergent (L-NRB) and assayed by measuring light output in the presence of luciferase-luciferin with a luminometer. The technique can be performed in about 45 min and bacterial numbers down to about 1×10^6 cfu/ml can be reliably enumerated.[13,21,23,28,59,61,86] Workers who have tried to correlate the ATP test with bacterial counts in the range 1×10^5 to 1×10^7 have concluded that, because of the high number of counts outside the 95% confidence limits, it may have limited use as a rapid microbiological test.[58] Other workers have commented more favorably[21,59] and, indeed, the test is being used successfully by dairies in some European countries.

Modifications to the original method have been made which claim to improve the performance in one way or another. For example, increasing the concentration of apyrase in the nonbacterial ATP extraction mixture enabled results to be achieved in 5 min with little change in sensitivity.[13,22,23] Other modifications include the incorporation of a protease[19] and a different sequestrant[23] in the nonbacterial ATP extractant to aid release of micellar-bound nucleotide. Alterations in reaction volume have reduced quenching and increased precision[13,23] but had little effect on sensitivity. Other bacterial extractants have been tried, including boiling Tris buffer[25] and butanol/octanol.[1,87] Poor results were achieved with the latter[87] and it was uncertain what, if any, precautions were taken to remove somatic cell ATP.

Incubation at 30°C for 30 min increased the intracellular ATP pool of *Enterobacter cloacae* from 0.62 to 1.06 fg per cell.[76] Following this temperature activation the precision of the bacterial ATP assay in milk also improved. However, the improvement in precision at low bacterial numbers (7.4×10^4) was not as great as that observed at higher population densities. Similarly, intracellular ATP concentrations did not rise to the expected levels after heat activation of these lower bacterial numbers. One major drawback to this approach of improving the test methodology is the extra time required.

An alternative way of correcting for somatic cell ATP in milk has been suggested by Langeveld and Van der Waals.[61] In their protocol, based on a technique devised for enumerating bacteria in meat,[88] two ATP determinations were carried out on the same milk sample. A sample of milk was assayed using the ATP platform test of Bossuyt.[22] An identical sample was treated in exactly the same way, but, instead of the addition of L-NRB to release bacterial ATP, an equivalent volume of water was added. The authors claimed that the difference in the light output obtained in the presence of the extractant and that found when water was used would be due to microbial ATP only. By this method, bacterial counts above 2×10^5 could be detected, compared with a value of 2×10^6 when the platform test was used. The standard deviation of the estimate was, however, larger for the difference method than the original platform test (0.41 and 0.25 log/ml, respectively).

2. Methods Involving Concentration of Bacteria

There are two main advantages to be gained if bacteria can be removed from milk prior to the assay of their nucleotide content. Firstly, the problems encountered due to high levels of nonbacterial ATP can be eliminated and, secondly, it allows concentration of the bacteria so that greater numbers can be presented to the assay system. This should result in increased sensitivity.

a. Concentration by Filtration

A method for rendering milk filterable has been described.[31] As outlined previously, it involves pretreatment of milk at 50°C in the presence of a protease and nonionic detergent. This enables the milk to be filtered through a bacteria-retaining membrane. Webster et al.[13] have optimized the filtration technique for bioluminescence. Milk (50 ml) was treated at 50°C for 10 min with Lubrol PX (8 ml of 0.5% solution) and trypsin (2 ml of 20% solution). They claimed that this pretreatment enabled filtration through a polycarbonate membrane filter (47 mm diameter; 0.4 μm pore size) aided by creation of a vacuum. Nonbacterial ATP was also removed during the pretreatment and subsequent filtration. The intracellular ATP was released from bacteria trapped on the filter surface by elution with NRB (1 ml). When aliquots of the eluted ATP were assayed with luciferase-luciferin, bacterial numbers in the original milk corresponding to 1×10^5 cfu/ml could be detected. Attempts to repeat this work[29] have failed to achieve filtration of the volumes of milk quoted in the paper. Also, the increase in sensitivity obtained was disappointing considering the initial volume of milk concentrated. The differences in volume of milk filtered did not seem to have the expected

TABLE 4
Zeta Potentials of Bacteria at Different pH

Organism	Zeta potential (mV)		
	pH 3	**pH 5**	**pH 10**
Pseudomonas diminuta	0	−30	−50
Serratia marcescens	−8	−22	−32

effect on sensitivity. For instance, when 50 ml of milk was filtered, the lowest number of bacteria which could be reliably detected was about 8×10^4 cfu/ml, but when only 10 ml was filtered this figure increased to only about 1×10^5 cfu/ml instead of the 4×10^5 cfu/ml expected. The background ATP levels found in the NRB extract of filtered cells appeared higher than those assayed by a modification of the Bossuyt assay,[21] which may suggest incomplete removal of nonbacterial ATP during the filtration procedure and accompanying pretreatment.

Another method for estimating the bacteriological quality of raw milk by filtration has recently been described.[64] The method consisted of mixing equal volumes (5 ml) of milk and an unidentified reagent (designated "NRS for filtration") at 37°C for 4 min. The reagent also contained apyrase at a concentration (2 U/ml) four times greater than that used by Bossuyt.[22] This treatment was designed to remove nonbacterial ATP and make the milk filterable. The treated sample was vacuum filtered through an N_{66} Posidyne Nylon 66 membrane (47 mm diameter; 0.65 μm pore size supplied by Pall Process Filtration Ltd). Extraction of the ATP from the bacteria retained on the membrane was achieved with NRM and assayed with luciferase-luciferin (Lumit QM). A detection limit of 2×10^4 cfu/ml was obtained. The good sensitivity of the assay coupled with its speed (7 min) and accuracy make the test of great potential use to the dairy industry. A filtration method was also described by Pernelle.[89] A sensitivity of at least 1×10^5 cfu/ml was achieved, although the nature of the ATP extractants and filtration procedure was uncertain.

The membranes used by Waes et al.[64] contain cationic (positively charged) functional groups that aid in the removal of bacteria which generally possess a net negative surface charge.[90] Typical Zeta potentials of a number of microorganisms are shown in Table 4. Positively charged filters have been used successfully to remove bacteria from milk.[91] There was some speculation that bacteria present in milk were removed by physical entrapment in the matrix of these filters and not by electrostatic interaction.[92] The Zeta Plus filters used by Kroll[91] were fibrous and bacteria were retained when milk was filtered through these. However, when milk was filtered through positively charged nylon membranes of the same nominal pore size (5 μm) there was no evidence of bacterial retention.[92]

A new luminometer, BactoFoss, is available that automatically performs a filtration step prior to extracting and assaying ATP from bacterial cells.[82,93] The instrument automatically removes a known volume of sample and deposits it in a temperature-controlled funnel where the milk is treated to remove somatic cell ATP and filtered. The bacteria remain on a filter paper belt that rotates to the detection chamber where the bacterial ATP is extracted and reacted with luciferase-luciferin. A linear relation with plate count down to 1×10^4 cfu/ml was found for 304 raw milk samples.[82] The regression line had a slope close to unity and a small standard deviation of estimate that were thought to be indicative of complete removal of nonbacterial ATP. This luminometer has obvious advantages when analyzing raw milk samples and should be of great benefit to the industry if the results claimed by the manufacturers can be confirmed. Results obtained with the original method[22] and with filtration methods[13,82] are shown in Figure 1.

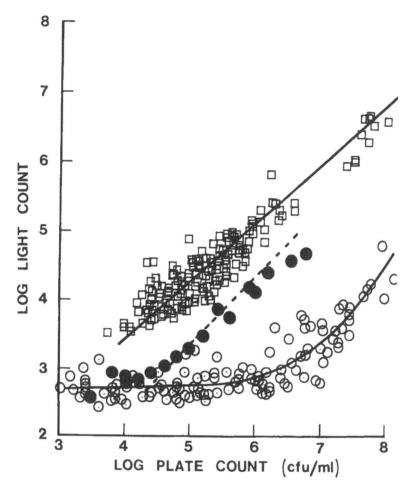

FIGURE 1. Estimation of bacterial count in raw milk by bioluminescence using (1) the modification of Bossuyt's method[22] described by Griffiths and Phillips[23] (O), (2) the data of Eriksen and Olsen obtained with the BactoFoss machine[80] (□), and (3) the scattergram of mean log plate count and mean log RLU for the filtration method as described by Waes et al.[63] (●).

b. Concentration by Other Methods

Because microorganisms are generally negatively charged at pH values above 5,[94] they will adsorb to positively charged matrices such as charged filters,[91] ion exchange resins,[95-97] and magnetite (the magnetic oxide of iron).[98,99] Griffiths and Phillips[23] reported that bacteria could be removed from suspension by magnetite and Celite (a diatomaceous silica available commercially). Magnetite was only of limited use for removing bacteria from milk as no *Pseudomonas* spp. tested were adsorbed. Celite and Zeta plus filters were equally effective in the removal of organisms from suspension in synthetic media, but the filters were more effective with milk. However, the physical nature of the Zeta plus filters made them of little value in conjunction with ATP methodology. Reasonable results were achieved when the bacterial ATP of cells removed by Celite was compared with the plate count of the original milk sample. This assay procedure appeared to be able to detect bacteria in milk at levels down to about 1×10^4 cfu/ml in approximately 15 min. The bacterial cells were probably removed from suspension by physical entrapment rather than electrostatic interaction with the matrix.

Bacteria also possess cell surface properties that confer hydrophobicity and it may be possible to separate microorganisms by hydrophobic interaction chromatography.[97]

Lectins are proteins which selectively bind carbohydrates but do not exhibit enzyme activity. They combine noncovalently with mono- and oligosaccharides in the same way that antibodies bind antigens.[100] A number of studies have been carried out on lectin-microbe interactions, including some species commonly found in milk.[101] Griffiths and Phillips[23] were able to effectively remove bacteria from milk with an immobilized lectin, Concanavalin A, specific for D-mannose and D-glucose residues.[102] Subsequent detection of the lectin-bound bacteria by the ATP technique gave a significant correlation with total bacterial counts in milks in the range 3×10^3 to 3×10^5.

Other methods of removing bacteria from foodstuffs may be useful in combination with ATP technology. These could include antibodies of different specificities to remove either specific bacteria[97] or groups of organisms such as *Pseudomonas* spp.

The development of reliable methods to remove and concentrate bacteria from milk has now reduced the sensitivity of the ATP test for milk bacteria to levels which make it appropriate for routine testing of raw milk at creameries.

III. ESTIMATION OF PASTEURIZED MILK QUALITY USING ATP TECHNOLOGY

The increasing importance of the supermarket as a retail outlet for pasteurized milk has led to pressure on processors to manufacture product with a longer shelf-life. Arguably, the best way of achieving this extension of keeping quality is to reduce the levels of bacteria reintroduced into product after pasteurization.[103,104] Other important factors governing shelf-life, such as maintenance of the cold chain during marketing, are to some extent out of the control of the processor. Postpasteurization contamination is also important from a public health standpoint. A number of large outbreaks of food poisoning associated with the consumption of pasteurized milk, including cases of listeriosis and salmonellosis, have been the direct result of post-heat-treatment contamination.[104] Concerns about the hygienic quality of pasteurized milk within the European Community led to the issue of a directive detailing microbiological standards for this product which have since been adopted in the U.K.[4] The principal test of keeping quality in the new regulations involves incubation of pasteurized milk at 6°C for 5 days, followed by a plate count at 21°C for 25 h, after which the count should not exceed 1×10^5 cfu/ml. Clearly, the results obtained by this test are of historical interest only. A much more rapid indication of the quality of pasteurized dairy products is required by the industry.

A. METHODS TO DETECT PRODUCT QUALITY

The postpasteurization contaminants that have the greatest effect on shelf-life are Gram negative psychrotrophic bacteria, principally of the genus *Pseudomonadaceae*.[103] These bacteria when present in milk even at very low numbers (ca. 1/100 ml) can have a significant effect on shelf-life.[105] Detection of bacteria at such low levels is difficult even with conventional tests and is impossible using current ATP technology. In order to raise bacterial numbers to detectable levels a sample of the pasteurized product can be pre-incubated at an elevated temperature for a short time period. Gram negative bacteria can be selected for by incorporating an inhibitor system to prevent Gram positive bacterial growth. The latter are able to grow at the pre-incubation temperatures usually employed, but generally not at refrigeration temperatures.[106] The procedures adopted over the years have been reviewed.[103,107] Waes and Bossuyt[108] described a method to detect postpasteurization contamination in pasteurized milks based on bacterial ATP measurement (BC-ATP test). Milk was pre-incubated at 30°C for 24 h in the presence of benzalkon A 50% (0.06%) and crystal violet (0.002%) to prevent the growth of Gram positive bacteria.[109] After this pre-incubation, the bacterial ATP content of the milk was assessed by Bossuyt's method.[21] A photometer reading of 1000 relative light units, that corresponded to a bacterial count of 6.5×10^5 cfu/

ml, was indicative of initial contamination levels of about 1/l of milk. The test could be made more quantitative by applying a most probable number approach.[110] In this case, 1000, 100, 10 and 1 ml quantities of milk containing the inhibitor solution were pre-incubated prior to extraction and assay of bacterial ATP. By performing the test on 100 ml milk samples, it could predict with 92% accuracy if a count of 1×10^5 cfu/ml was reached after storage of the original milk at 7°C for 5 days. The discrepancies were due to the insensitivity of the BC-ATP test, the variable growth rate of Gram negative bacteria present, or spoilage due to Gram positive bacteria such as *Bacillus* spp.

The pre-incubation conditions used by Waes and Bossuyt[108,110] were probably not ideal for selection of Gram negative psychrotrophs. The inhibitors to prevent Gram positive bacterial growth were also inhibitory toward a number of Gram negative bacteria.[108,109,111] A better indication of levels of postpasteurization contamination was obtained by a method involving the use of less aggressive inhibitors.[108] In addition, a pre-incubation temperature of 30°C would encourage growth of mesophilic bacteria that do not grow at refrigeration temperatures and true psychrotrophic bacteria would not grow optimally.[112] Rodrigues and Pettipher[113] obtained a better correlation with keeping quality and plate count after incubation of pasteurized milks at 20°C rather than 30°C. Thus, the pre-incubation procedure would not accurately reflect numbers of postpasteurization contaminants present in the original milk sample. Phillips et al.[106] proposed different pre-incubation conditions that were thought to be more selective for the types of Gram negative psychrotrophs reintroduced into milk after the heat treatment. In their test (P-INC test), product containing an inhibitor solution comprised of crystal violet, penicillin, and nisin at final concentrations of 20 μg/ml, 200 U/ml and 400 U/ml, respectively, was incubated at 21°C for 25 h. This time-temperature combination had previously been shown to produce the same level of growth as expected during incubation at 6°C for 10 to 14 days.[114,115] The mixture of crystal violet, penicillin, and nisin (CPN) was less inhibitory toward Gram negative psychrotrophic bacteria than benzalkon-crystal violet.[111] The numbers of bacteria present in pasteurized milk and creams after this P-INC test gave a good indication of initial levels of postpasteurization contamination.[106,116] The bacteria present in the pre-incubated sample could be enumerated by measuring their ATP content.[117,118] The P-INC test coupled with ATP assay correlated well with results achieved using plate counts (95.4% agreement). The accuracy of prediction of quality was only slightly lower (94.5%) when P-INC ATP counts were compared with results obtained with a shelf-life test that consisted of enumerating bacteria in the product after storage at 6°C for 7 days. The shelf-life of pasteurized milk,[119,120] single,[119] and double cream[119] could be predicted by the P-INC ATP test (Figure 2). Shelf-life in this study was defined as the time taken for the bacterial count in the product to reach 1×10^7 cfu/g upon storage at 6°C. This had earlier been shown to correlate well with shelf-life determined organoleptically. Nevertheless, some criticism has been made of this method of measuring keeping quality.[6] The relation between shelf-life of pasteurized milk and pre-incubation count obtained by bacterial ATP assay was significantly better when CPN and not benzalkon-crystal violet was used as the inhibitor system.[119] However, there was no significant difference between the inhibitors when pasteurized single cream was pre-incubated.

Both the BC-ATP[108,109] and the P-INC ATP[117-120] tests allow prediction of levels of postpasteurization contamination and, hence, shelf-life in about 25 to 26 h after processing, giving a significant saving in time over other methods. The P-INC test procedure also gave good agreement with the EC keeping quality test required by new legislation.[121]

B. SUITABILITY OF PLATE COUNT AS COMPARISON FOR BACTERIAL ATP ASSAY

Arguments have raged since the first ATP measurements were made on foods about the suitability of plate counts as a reference method.[1,6,68,122] Problems may result from the

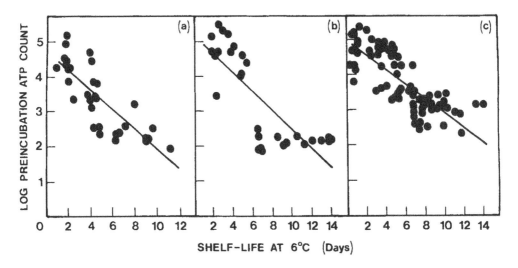

FIGURE 2. Comparison of bacterial ATP concentration of sample after pre-incubation at 21°C for 25 h in the presence of crystal violet-penicillin-nisin with shelf-life at 6°C for (a) pasteurized milk, (b) pasteurized single cream, and (c) pasteurized double cream. ATP concentration is expressed as log relative light units. (From Phillips, J. D. and Griffiths, M. W., *Food Microbiol.*, 2, 39, 1985. With permission.)

clumping of cells, making it difficult to assess the numbers of bacteria that gave rise to a single colony. The medium used for counting may not be suitable for resuscitation of sublethally injured cells. Bacterial ATP concentrations in food reflect the metabolic activity of those organisms and may be more useful in assessing spoilage potential or shelf-life.[6] LaRocco et al.[68] suggested that a more appropriate reference method for comparison and interpretation of ATP results would be an assay that measures cellular metabolic activity. One method that was considered of potential benefit was impedance monitoring. Bossuyt and Waes[123] showed there was a relation between impedance detection time and bacterial ATP concentrations in pre-incubated pasteurized milk, but the ATP data were presented in a most probable number form and not in absolute terms. Re-examination of results obtained in my laboratory showed a good relation between ATP values and detection times for pre-incubated pasteurized milk. However, the correlation was no better than that obtained when ATP readings were compared with plate counts on the same samples. While being far from perfect, in the absence of anything better, plate counts will remain the method by which all others are compared.

C. HYGIENE MONITORING OF PROCESSING PLANT

In order to be able to respond rapidly to problems arising from contamination of processing plant, or even to detect inadequate processing, management must have a rapid system for indicating hygiene failures. The P-INC ATP test can be used, in conjunction with in-line sampling, to detect relatively rapidly areas in the plant that have been inadequately cleaned. Remedial measures can then be taken to ensure product quality and safety. Significant improvements to the quality of pasteurized milk and cream were made when these methods were adopted by processors.[120]

Swabbing has long been used in the food industry to assess the cleanliness of surfaces. The bacterial load on the swab has traditionally been determined by plate count. However, the total ATP present on the swab can be extracted and assayed by bioluminescence.[6] The concentration of ATP is directly proportional to the level of contamination on the original surface swabbed as even the presence of nonmicrobial ATP is indicative of the presence of organic debris. An estimate of plant cleanliness can be obtained in less than 5 min using this technique. The rapidity of the test, coupled with the availability of portable luminometers, suggest that it may have widespread application in food factories including creameries.

IV. STERILITY TESTING OF UHT PRODUCTS

Because of the high heat treatment (ca. 140°C for 1 to 5 s) required for manufacture of UHT products, they are essentially sterile. Specially designed fillers for aseptic packaging of these products ensure that sterility is maintained throughout production. Very low levels of contamination are, therefore, encountered and usually fewer than one in 1000 packs are expected to be nonsterile. This requires relatively large samples to be tested to ensure the quality of a production batch. A rapid, simple, reliable test is desirable to facilitate handling of large numbers of samples. A rapid test of sterility would have the added advantage that product could be released from the factory sooner, thus saving on warehouse costs.

A. COMPARISON OF STERILITY TEST METHODS

Sterility testing of UHT products is normally carried out after incubation at 30°C for 3 to 5 days. After this storage period, microbial growth in the product is determined by plate count or by monitoring pH changes. Incubation followed by plate count is a laborious and time-consuming exercise. Results are not available for 5 to 8 days. Whereas results are obtained more rapidly by following pH changes during the incubation, they are often unreliable, especially when contamination is due to sporeformers.[124] An alternative method of monitoring growth during storage of the product at 30°C was proposed by Waes et al.[124] Samples of UHT milk were treated with apyrase (0.01 U) prior to extraction of bacterial ATP with L-NRB. The light emission after reaction with luciferase-luciferin was noted. If this was below 1000 relative light units after 3 days' storage of the milk at 30°C it was considered sterile. There was no need for extraction of somatic cell ATP in UHT milks as they would not be expected to remain intact after the heat treatment.[16,36,40] A number of 1 l UHT packs was sampled during replacement of the paper roll on the filling machine. At this time considerable contamination of the milk could be expected. After 3 days' incubation at 30°C, 84% of the samples were identified as being nonsterile by plate count, compared with 75% by the ATP test. When the storage period was extended to 9 days there was complete agreement between the plate count and ATP detection methods. All the contamination caused when UHT milk was understerilized could be detected in 3 days by the ATP method.[124] Other work[125] has shown significant detection of contamination using a 1 day pre-incubation at 30°C followed by bacterial ATP determination. Extension of the pre-incubation period to 3 days gave results with 100% confidence.

Quesneau[69] concluded that sterility testing of UHT milks by bioluminescence was no more sensitive than a method involving dye reduction. Milk contaminated at levels of 1.8 × 10³ cfu/ml could be detected after 4 h incubation at 37° while samples containing 10 cfu/ml required 20 h incubation. Bioluminescence was useful for sterility testing of chocolate milks where dye reduction could not be used.

B. DETECTION OF MILK-DEGRADING ENZYMES BY BIOLUMINESCENCE

As a consequence of the growth of most psychrotrophic bacteria in milk and dairy products, extracellular enzymes are produced that can degrade milk protein and fat.[104,126-130] These proteases and lipases are often extremely heat stable and can retain about 60% of their activity after pasteurization and some 40% of activity after UHT.[131] Thus, relatively small amounts of enzyme present in raw milk may result in degradation of the heat-treated product. This is especially true in the case of UHT products when storage for long periods at ambient temperatures occurs. Simple, rapid methods for detection of low concentrations of these enzymes would be of benefit in assessing the suitability of raw milk for UHT processing and even for testing UHT products themselves.

1. Proteases

The effect of protease in raw milk on the shelf-life of UHT milk has been demonstrated.[132,133] The storage life of UHT milk produced from raw milk heavily contaminated with a protease-producing psychrotroph was only about 25% of that obtained with controls made from uninoculated milk.[132] As no defined relation exists between bacterial count and protease level,[134] a direct measurement of the proteolytic activity has to be obtained for quality control of sterilized products.[135] A comparison of a number of methods proposed for the assay of bacterial proteases in milk revealed there was no simple rapid test available.[135] Those most suitable involved appreciable sample preparation, the use of hazardous chemicals, and long incubation periods. A protease assay based on bioluminescence has been described.[136,137] The rate of inactivation of bacterial[136] or firefly[137] luciferase due to proteolysis was directly proportional to the protease concentration at a constant initial luciferase level. Griffiths[29] used firefly luciferase as the substrate for the protease and was able to show that trypsin activity could be measured in sterile reconstituted skim milk at levels of at least 4 mU/ml (Figure 3). This is well within the sensitivity required, as it has been determined that the threshold value for enzyme activity to achieve a shelf-life for UHT milk in excess of 5 months at 23°C was 30 mU/ml.[138] Current research suggests that an assay based on this methodology may be able to detect similar levels of bacterial protease in milk within 1 h.[139]

The assay system has the advantage of being unhindered by contaminating materials as well as being rapid, sensitive, and simple. However, care must be exercized to ensure assay conditions that do not result in inactivation of luciferase in the absence of protease and that do not protect luciferase from protease attack.

2. Lipases

Mottar[140] was able to demonstrate a relation between psychrotrophic counts of raw milk and residual enzyme activity in UHT milks manufactured from it. When lipase from *P. fluorescens* was added to UHT milk at levels commonly found in raw milk (0.3 U/ml), the heat-treated milk became rancid within 5 to 8 days storage at 8°C. Assay methods able to detect these levels quickly and with the minimum of sample preparation are required. A common method of detecting lipolysis in milk products relies on determining levels of free fatty acids produced during the reaction.[130] This is usually achieved by extraction and subsequent titration. A simple, highly specific, and sensitive bioluminescent method for determination of free fatty acids has been developed.[141] Free fatty acids present are activated by acyl CoA synthase and coupled by a series of enzyme reactions (Figure 4) to the bacterial NADH-linked luciferase. This assay system has not, to the reviewers' knowledge, been applied to milk.

An elegant bioassay for the detection of lipase has been described.[142-144] This utilizes a ''dim'' mutant of a luminous bacterium that will emit light only in the presence of long chain fatty acids, especially myristic acid.[145] The assay is represented schematically in Figure 4. The luminescence response is proportional to the amount of added myristic acid. Using trimyristin as a substrate, lipase activity can be detected, as well as phospholipases A_2 and C.[144] Again, this system has not been tested fully in milk but has considerable potential.

V. MONITORING THE ACTIVITY OF STARTER CULTURES

A. BIOLUMINESCENCE TO MONITOR STARTER ACTIVITY

The manufacture of fermented milk products including cheese and yogurt relies on the production of lactic acid as a result of the metabolic activity of microorganisms in milk. These organisms are usually referred to as starter cultures. The production of acid by starter cultures in milk can be prevented by a number of factors. By far the most important are the

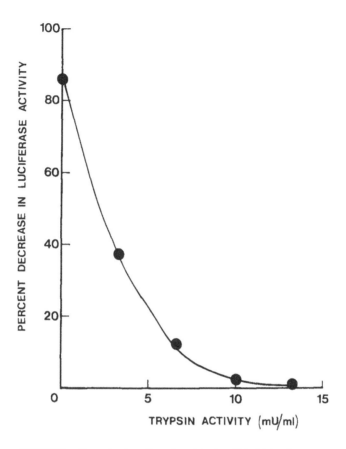

FIGURE 3. Determination of trypsin activity in milk by bioluminescence. The decrease in luciferase activity in the presence of varying levels of trypsin was assessed after 4 h incubation at 30°C against a standard ATP solution.

presence of antibiotic residues in the milk introduced during therapy for mastitic infections and the presence of bacteriophage that can lyse the lactic acid-producing bacteria. There is a strong correlation between acid production and ATP levels for two strains of *Lactococcus lactis* and one strain of *Lactobacillus acidophilus* grown in sterile, homogenized milk.[146] A similar relation was found for an unnamed mesophilic starter organism and a *Lactococcus* yogurt culture.[147] It has been proposed[148] that highly active yogurt starter cultures showed the greatest increase in ATP levels within 3 min of inoculation into milk at 37°C, although 1 to 2 h may be a more realistic response time. By comparing curves of increase in ATP with time produced for the starter bacterium in sterile milk and in the production milk, an indication of the presence of inhibitory substances can be obtained. No account needed to be taken of the somatic cell ATP, as the levels of bacterial ATP in the milk were determined from difference measurements. It was concluded that lower concentrations of inhibitors present in milk could be detected by incorporation of apyrase in the assay system.[149] As well as providing an indication of the presence of inhibitory substances in the milk for fermentation, ATP assay may provide a better indication of the metabolic status of the starter cultures than obtained by measuring pH changes.

Numerous studies have been performed on bioluminescence methods to detect antibiotics in milk.[89,147,149-151] The organisms used in the susceptibility tests included strains of the lactic acid-producing bacterium, *Lactococcus thermophilus*.[89,147,150,151] By monitoring variations in ATP concentrations during growth of the culture in milk, at least 0.005 U/ml of penicillin could be detected in about 90 min[147,151] (Figure 5).

BASIS OF BIOLUMINESCENT LIPASE ASSAY

a)   "DIM" mutants of the luminous bacteria *Beneckea harveyi* emit light only in the presence of exogenous aldehydes or fatty acids.

$$\text{WASHED CELLS + SUBSTRATE + LIPASE SOURCE} \longrightarrow \text{LIGHT}$$
$$\text{(trimyristoyl}$$
$$\text{glycerol)}$$

b)   By determination of free fatty acids

1)   FFA + CoA + ATP $\leftrightarrow$ AcylCoA + AMP + PPi
        acylCoA synthetase

2)   PPi + FRU-6-P $\leftrightarrow$ FRU-1,6-DIPHOS + Pi
        pyrophosphate fructose-6 phosphate phosphotransferase

3)   FRU-1,6-DIPHOS $\leftrightarrow$ DHAP + GLY-3-P
                aldolase
                $\uparrow\_\_\_\_\_\uparrow$
                triosephosphate isomerase

4)   GLY-3-P + NAD$^+$ + AsO$_4^{3-}$ $\rightarrow$ 3PG + NADH + H$^+$ + HAsO$_4^{3-}$
        G-3-P^A dehydrogenase

5)   NADH + H$^+$ + FMN $\rightarrow$ FMNH$_2$ + NAD$^+$
        oxidoreductase

6)   FMNH$_2$ + R-CHO + O$_2$ $\rightarrow$ R-COOH + FMN + LIGHT
        luciferase

FIGURE 4.   Schematic representation of bioluminescent lipase assay using (a) ''dim'' mutants of a luminous bacteria,[140-143] and (b) an enzyme cascade reaction for determination of free fatty acid levels.[139]

B. NOVEL METHODS TO DETECT BACTERIOPHAGE

An immunochemiluminescent assay for the detection of different types of phages involved in the lysis of lactic acid-producing streptococci has been developed.[152]

An alternative approach involves the use of genetically engineered starter organisms with a bioluminescent phenotype. A marine bacterium, *Vibrio fischeri*, possesses a luciferase enzyme.[153] The light-emitting reaction differs from the firefly luciferase as it involves the oxidation of FMNH$_2$ together with a long chain aliphatic aldehyde. For a nonluminous bacterium to become bioluminescent, the genetic transfer of the luciferase and fatty acid reductase genes (*lux* genes) is required. These genes have been cloned and successfully transferred to a number of bacteria.[154-156] Researchers at Nottingham University have engineered a plasmid-based replicon termed pSB154, for introduction of luciferase into a wide range of lactic acid bacteria including *Lactococcus lactis*, *Lactobacillus casei* and *Lactobacillus plantarum*.[152] The long chain aldehyde can be added exogenously to provide a bioluminescent phenotype. Because cell lysis can be easily measured by following the decay of luminescence,[157] these genetically engineered lactic acid bacteria would be ideally suited as indicators of the presence of lytic phages and antibiotics.

VI. DETECTION OF BACTERIA IN OTHER DAIRY PRODUCTS

Bioluminescence techniques have not been widely used to determine the bacterial content of products other than those mentioned earlier in this chapter. McMurdo and Whyard[158]

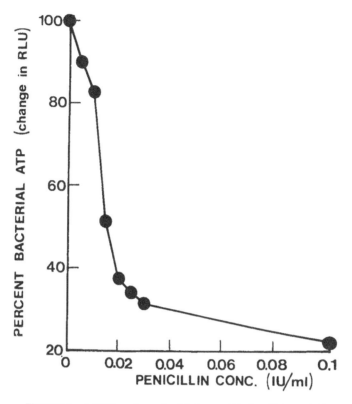

FIGURE 5.   Inhibition of growth of *S. thermophilus* in milk by penicillin. Growth was determined by bacterial ATP assay after 90 min at 45°C. (Data from Williams, G. R., *J. Soc. Dairy Technol.*, 37, 40, 1984. With permission.)

did investigate a bioluminescent method to detect bacterial contamination of milk powders. The levels of microbial ATP in dried milk were below the detection limit of the assay system. One reason suggested for this was that the normal flora of the milk powder consisted largely of *Bacillus* spp. present as endospores. It is known that spores contain little ATP[159] and this may have accounted for the low ATP levels observed in the powders. To overcome this problem the reconstituted milk powder was incubated for 5 h at 30°C to increase microbial numbers. After extraction and hydrolysis of somatic cell ATP, which was entirely unnecessary as this would be negligible in milk powder, the bacterial ATP was assayed in the usual manner. An internal standard was added to allow for quenching. However, the manner in which this standardization was carried out was unreliable. Nevertheless, a correlation was found between ATP concentration and plate count for the milk powder samples at counts in the range 2×10^2 to 1×10^5 cfu/ml. The degree of scatter about the regression line was considered unacceptable. Some possible sources of error were listed and included error in the timing of the enrichment and destruction of nonmicrobial ATP, errors in preparation and pipetting of sample and ATP standard, deterioration of the standard during use, and variations in the flora after enrichment. The latter was considered liable to introduce errors resulting from species specific variations in ATP content of cells. The true reason for the errors encountered was probably the low numbers of bacteria found in even the incubated samples. Better results would have been achieved if the incubation had been extended to allow bacterial populations to reach 1×10^5 cfu/ml and over.

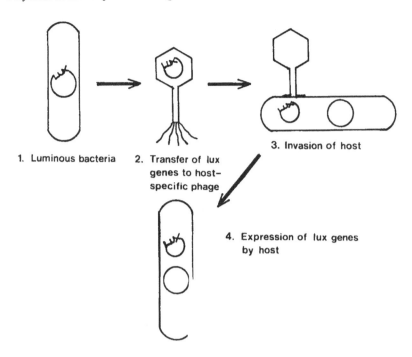

1. Luminous bacteria   2. Transfer of lux
                           genes to host-
                           specific phage

3. Invasion of host

4. Expression of lux genes
   by host

FIGURE 6.   Schematic representation of a method for detecting specific groups of bacteria by *lux* gene technology.

VII. DETECTION OF SPECIFIC GROUPS OF BACTERIA

Recent concerns about the presence of pathogens, particularly *Listeria monocytogenes*,[160] in dairy products has highlighted the need for simple, reliable, rapid test systems for the identification of these organisms. Advances in molecular biology have meant that detection systems for specific types of bacteria based on bioluminescence are now a real possibility.

A. *LUX* GENE TECHNOLOGY

Each bacterial genus has its own associated bacteriophages that vary in host specificity. As mentioned earlier in this review, the genes responsible for bacterial luminescence have been identified and cloned.[154-156] The DNA carrying these genes can be introduced into bacteriophages and will be transferred into the host bacterium during infection.[154] Once inserted into the host, transcription of the *lux* genes results in light emission that can easily be detected with a luminometer. Expression of the genes is entirely dependent on host bacterial metabolism as the phage possesses no metabolic activity. The magnitude of the light emission is directly related to the number of cells infected and the specificity of the assay is governed by the specificity of the phage. Therefore, bacteria susceptible to "luminescent" phage infection will be detected in the presence of other noninfected bacteria. A schematic representation of the procedure is given in Figure 6. A bacteriophage, P22, essentially specific for *Salmonella typhimurium* has been genetically engineered to contain the *lux* genes. Infection of a culture containing *S. typhimurium* by the phage allowed detection of as few as 1×10^2 bacteria[152] (Figure 7). An astonishing sensitivity of ten cells per ml within 1 h has been claimed when the assay has been applied to the detection of *E. coli* in milk.[161]

By introducing *lux* genes into phages specific for enteric bacteria[152] or even for *Pseudomonas* spp., an on-line hygiene test for dairy products becomes a possibility. Stewart et al.[152] conceive that such a test would take no more than 1 h to perform and would not require major capital expenditure. Furthermore, it would enable factory personnel to obtain

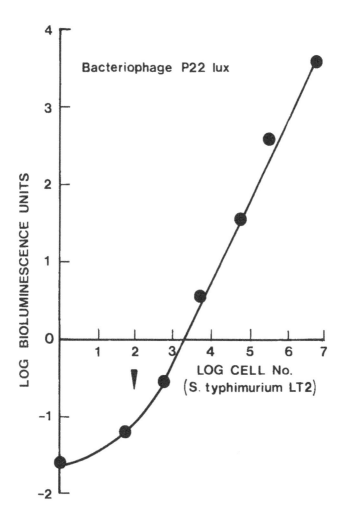

FIGURE 7. Detection of *S. typhimurium* LT2 by infection with a *lux*+ derivative of bacteriophage P22. Bioluminescence was measured 60 min after infection. The arrow indicates where the instrument background and sensitivity determine the limit of detection. (From Stewart, G., Smith, T., and Denyer, S., *Food Sci. Technol. Today*, 3 (1), 19, 1989. With permission.)

an accurate assessment of the levels of enteric bacteria, or other organisms indicative of poor hygiene, in raw materials, product during manufacture, end product, and environmental samples. They point out the obvious advantages of combining such a test with the Hazard Analysis Critical Control Point (HACCP)[162] approach to quality assurance.

B. OTHER SELECTIVE METHODS

1. Differential Growth

Media that are selective for a specific group of bacteria have formed the basis of identification procedures in microbiology since the earliest days of the science. Growth of bacteria in these selective media can be detected by bioluminescence. In fact a test based on this principle for the detection of Gram negative psychrotrophic bacteria has already been described in this review. An ATP-free medium for the differential growth of coliform organisms has also been described.[158] Such selective methods have been applied for the detection of coliforms,[163] *Salmonella* spp.[164-166] and, more recently, *Listeria* spp.[167] using

TABLE 5
Extractant Concentration for Maximum ATP
Yields with Different Microorganisms[a]

Microorganism	Extractant conc. (%w/v) for maximum ATP yield		
	TCA[b]	PCA[c]	DTAB[d]
Bacillus cereus	5.0	1.25	1.0
Staphylococcus aureus	2.5	1.25	0.5
Klebsiella pneumoniae	2.5	0.16	1.0
Pseudomonas aeruginosa	0.63	0.16	1.0
Escherichia coli	1.25	0.31	1.0
Candida albicans	5.0	2.5	[e]
Scenedesmus obtusiusculus	10.0	5.0	[e]

[a] Data of Lundin.[50]
[b] Trichloroacetic acid.
[c] Perchloric acid.
[d] Dodecyl trimethyl ammonium bromide.
[e] Extractant unable to lyse cells.

impedance monitoring to identify growth. An inhibitor system supposedly selective for *Pseudomonas* spp. has also been assessed against a number of bacteria of dairy origin.[111] Such media could be modified for use with bioluminescence techniques to detect growth. The main disadvantages of this approach are the inadequate specificity of most supposedly selective media and the incubation time required to obtain detectable numbers of bacteria. However, these techniques are relatively inexpensive and may provide a useful first step in a screening procedure for a particular organism.

2. Differential Extraction

Lundin[50] studied the effects of a number of nucleotide extractants on a variety of bacteria. Results showed (Table 5) that the concentration of extractant required to produce >75% and maximal yields of nucleotide was dependent on the bacterium. For example, of the bacteria tested, *Klebsiella pneumoniae* was most susceptible to perchloric acid but least affected by trichloroacetic acid. This may have been due to an artifact, such as increased ATP production before the cells lyse[50] or one of the many other factors that affect bacterial ATP levels.[53] Despite these criticisms it does illustrate the possibility for selective extraction of ATP from different bacterial types.

Tsai et al.[168] introduced an interesting concept of differential filtration followed by selective extraction of bacterial ATP. Filtration of a suspension of a mixed microbial population through membrane filters of decreasing pore size arranged in series allowed the separation of yeasts and molds from some bacteria. Levels of each type of microorganism could be detected by assaying ATP concentrations present on the appropriate filter. The same approach might be applicable for separation of large bacteria (e.g., *Bacillus cereus*) from those with smaller cells such as *E. coli*. Differentiation and detection of *Staphylococcus aureus* from other bacteria such as *E. coli* was achieved with lysostaphin. This is an agent that specifically lyses staphylococcal cells and the released ATP can be detected by luminometry. The efficiency of ATP extraction by lysostaphin was higher for *S. aureus* (100%) than *E. coli* (11%).[168] The use of other differential lytic agents such as lysozyme, lyticase, and, perhaps, various antibiotics has not been widely studied and deserves some attention.

VIII. CONCLUSIONS

In a recent survey of professional microbiologists[169] in the U.S. only 3.6% of those questioned were using bioluminescence techniques routinely. In a similar survey conducted in the U.K.,[122] bioluminescence had been evaluated by 50% of the respondents, but only one, from overseas, was using it routinely. The application for which it was being used was the assessment of total counts in raw milk and sterility testing of dairy products. Poor reagent stability, high cost per test, lack of automation, and low specificity were some reasons cited for nonadoption. Undoubtedly, the main factor limiting the use of bioluminescence in the dairy industry is its lack of sensitivity due to high background ATP levels. These problems are being addressed and filtration methods are now appearing that will give the desired performance. Improvements in instrumentation, especially automated machines specifically designed for the food industry that eliminate laborious sample preparation steps, will expand the interest in ATP methodology. The appearance of truly portable instruments allied with better bacterial extractants makes the use of bioluminescence for hygiene testing of food processing plants a very attractive proposition. Arguably the greatest impact will be from developments in *lux* gene technology which will allow very rapid and specific detection of low numbers of pathogenic or indicator organisms. This coupled, perhaps, with simple "dipstick" tests could revolutionize hygiene monitoring.

Cost was not identified as a major factor limiting the adoption of "new technology" in the survey by Fung et al.[169] While acknowledging that reagents for ATP assay are expensive in relation to conventional procedures, the savings in time more than compensate. In comparison to other techniques, including immunological methods and gene probes, ATP is relatively cheap. Furthermore, the possibility of producing genetically engineered luciferase and luciferin may lead to reduction in cost.

Rapidity of results is the main criterion used for choosing microbial testing systems.[169] This is closely followed by ease of use, and accuracy. Of all the commercially available technologies for enumerating bacteria, bioluminescence is the most rapid and the most versatile. The research currently being conducted may lead to it becoming the most widely used.

ADDENDUM

Recently a test for enumeration of bacteria in milk has become available that incorporates a novel somatic cell ATP extractant (Somex A) and an easy to use filtration procedure.[170] Using this Milk Microbial ATP Kit, Griffiths et al.[170] analyzed 240 raw milk samples and obtained a linear relation with plate count in the range 5.2×10^3 to 3.7×10^7 cfu/ml.

The bioluminescence assay for protease in milk has also been further developed[171] and levels of the enzyme as low as 0.01 U/ml can be detected in milk in 5 min. Confirmation of the usefulness of the hygiene monitoring procedure based on ATP bioluminescence in a cheese manufacturing site has been reported.[172]

The *lux* genes have been successfully cloned into *Lactobacillus casei* and *Lactococcus lactis*.[173]

REFERENCES

1. **Sharpe, A. N., Woodrow, M. N., and Jackson, A. K.,** Adenosinetri-phosphate (ATP) levels in foods contaminated by bacteria, *J. Appl. Bacteriol.,* 33, 758, 1970.
2. **Forrest, W. W.,** Adenosine triphosphate pool during the growth cycle in *Streptococcus faecalis, J. Bacteriol.,* 90, 1013, 1965.
3. **Griffiths, M. W.,** Microbial estimation in dairy products, in *ATP Luminescence: Rapid Methods Microbiology,* Society of Applied Bacteriology, Tech. Ser. No. 26, Stanley, P., McCarthy, B., and Smither, R., Eds., Blackwell, Oxford, 1989, 167.
4. **Anon.,** *The Milk (Special Designations) (Scotland) Order 1988 Statutory Instruments,* H.M.S.O., London, No. 2191 (S.213), 1988.
5. **Williams, M. L. B.,** The limitations of the Du Pont Luminescence Biometer in the microbiological analysis of foods, *Can. Inst. Food Technol.,* 4, 187, 1971.
6. **Stannard, C. J. and Gibbs, P. A.,** Rapid microbiology: applications of bioluminescence in the food industry, *J. Biolumin. Chemilumin.,* 1, 3, 1986.
7. **Kay, H. D. and Marshall, P. G.,** Phosphorous compounds of milk. IV. Presence of adenine nucleotide in milk, *Biochem. J.,* 22, 416, 1928.
8. **Richardson, T., McGann, T. C. A., and Kearney, R. D.,** Levels and locations of adenosine 5'-triphosphate in bovine milk, *J. Dairy Res.,* 47, 91, 1980.
9. **Brunner, J. R.,** Physical equilibria in milk: the lipid phase, in *Fundamentals of Dairy Chemistry,* 2nd ed., Webb, B. H., Johnson, A. H., and Alford, J. A., Eds., Avi, Westport, CT, 1974, 474.
10. **Johnson, A. H.,** The composition of milk, in *Fundamentals of Dairy Chemistry,* 2nd ed., Webb, B. H., Johnson, A. H., and Alford, J. A., Eds., Avi, Westport, CT, 1974, 1.
11. **McGann, T. C. A. and Pyne, G. T.,** The colloidal phosphate of milk. III. Nature of its association with casein, *J. Dairy Res.,* 27, 403, 1960.
12. **Zulak, I. M., Patton, S., and Hammerstedt, R. H.,** Adenosine triphosphate in milk, *J. Dairy Sci.,* 59, 1388, 1976.
13. **Webster, J. J., Hall, M. S., Rich, C. N., Gilliland, S. E., Ford, S. R., and Leach, F. R.,** Improved sensitivity of the bioluminescent determination of numbers of bacteria in milk samples, *J. Food Prot.,* 51, 949, 1988.
14. **Emanuelson, U., Olsson, T., Mattila, T., Astrom, G., and Holmberg, O.,** Effects of parity and stage of lactation on adenosine triphosphate, somatic cell count and antitrypsin content in cows' milk, *J. Dairy Res.,* 55, 49, 1988.
15. **Waldschmidt, M.,** Metabolite levels and enzyme activities in the bovine mammary gland at different stages of lactation. I. Metabolite levels related to energy production, *J. Dairy Res.,* 40, 7, 1973.
16. **Olsson, T., Sandstedt, K., Holmberg, O., and Thore, A.,** Extraction and determination of adenosine 5'-triphosphate in bovine milk by the firefly luciferase assay, *Biotech. Appl. Biochem.,* 8, 361, 1986.
17. **Wooding, R. B. P., Peaker, M., and Linzell, J. L.,** Theories of milk secretion: evidence from the electron microscope examination of milk, *Nature,* 226, 762, 1970.
18. **DeLuca, M.,** Firefly luciferase, in *Advances in Enzymology,* Vol. 44, Meister, A. M., Eds., John Wiley & Sons, New York, 1976, 37.
19. **Packard, R. A. and Marth, E. H.,** Modifying the ATP bacterial test for use with raw milk, *J. Dairy Sci.,* 66, (Suppl.1), 67, 1983.
20. **Olsson, T., Gulliksson, H., Palmeborn, M., Bergstrom, K., and Thore, A.,** Methodological aspects on the firefly luciferase assay of adenine nucleotides in whole blood and red blood cells, *Scand. J. Clin. Lab. Invest.,* 43, 657, 1983.
21. **Bossuyt, R.,** Determination of bacteriological quality of raw milk by an ATP assay technique, *Milchwissenschaft,* 36, 257, 1981.
22. **Bossuyt, R.,** A 5-minute ATP platform test for judging the bacteriological quality of raw milk, *Neth. Milk Dairy J.,* 36, 355, 1982.
23. **Griffiths, M. W. and Phillips, J. D.,** Rapid assessment of the bacterial content of milk by bioluminescent techniques, in *Rapid Methods for Foods, Beverages and Pharmaceuticals,* Society for Applied Bacteriology, Tech. Ser. No. 25, Stannard, C. J., Pettit, S. B., and Skinner. F. A., Blackwell, Oxford, 1989, 13.
24. **Lin, S. H. C., Dewan, R. K., Bloomfield, V. A., and Morr, C. V.,** Inelastic light scattering studies of the size distribution of bovine casein micelles, *Biochemistry,* 10, 4788, 1971.
25. **Theron, D. P., Prior, B. A., and Lategan, P. M.,** Determination of bacterial ATP levels in raw milk: selectivity of non-bacterial ATP hydrolysis, *J. Food Prot.,* 49, 4, 1986.
26. **Kuyper, A. C.,** The quantitative precipitation of citric acid, *J. Biol. Chem.,* 123, 405, 1938.
27. **Pyne, G. T. and McGann, T. C. A.,** The colloidal phosphate of milk. II. Influence of citrate, *J. Dairy Res.,* 27, 9, 1960.
28. **Neaves, P.,** Evaluation of the Lumac ATP assay for estimating the total, viable bacterial load of raw milk, Report No. 900, Milk Marketing Board Technical Division, Thames Ditton, 1985.

29. **Griffiths, M. W.**, Unpublished data, 1989.

30. **Peterkin, P. I. and Sharpe, A. N.**, Membrane filtration of dairy products for microbiological analysis, *Appl. Environ. Microbiol.*, 39, 1138, 1980.

31. **Pettipher, G. L., Mansell, R., McKinnon, C. H., and Cousins, C. M.**, Rapid membrane filtration-epifluorescent microscopy technique for direct enumeration of bacteria in raw milk, *Appl. Environ. Microbiol.*, 39, 423, 1980.

32. **Schram, E. and Weyens-van Witzenburg, A.**, Control of experimental factors involved in luminescent ATP assays, in *ATP Luminescence: Rapid Methods in Microbiology*, Society of Applied Bacteriology, Tech. Ser. No. 26, Stanley, P., McCarthy, B., and Smither, R., Eds., Blackwell, Oxford, 1989, 37.

33. **Chappelle, E. W., Picciolo, G. L., and Deming, J. W.**, Determination of bacterial content in fluids, in *Methods in Enzymology*, Vol. 57, DeLuca, M. A., Ed., Academic Press, New York, 1978, 65.

34. **Brolund, L.**, Cell counts in bovine milk. Causes of variation and applicability for diagnosis of subclinical mastitis, *Acta Vet. Scand.*, Suppl. 80, 1985.

35. **Harding, F.**, The impact of central testing on milk quality, *Dairy Ind. Int.*, 52(1), 17, 1987.

36. **Bossuyt, R.**, Usefulness of an ATP assay technique in evaluating the somatic cell content of milk, *Milchwissenschaft*, 33, 11, 1978.

37. **Emanuelson, U., Olsson, T., Holmberg, O., Hageltorn, M., Mattila, T., Nelson, L., and Astrom, G.**, Comparison of some screening tests for detecting mastitis, *J. Dairy Sci.*, 70, 880, 1987.

38. **Malkamaki, M., Mattila, T., and Sandholm, M.**, Bacterial growth in mastitic milk and whey, *J. Vet. Med. B*, 33, 174, 1986.

39. **Siro, M.-R.**, Monitoring microbial growth by bioluminescent ATP assay, in *Rapid Methods and Automation in Microbiology and Immunology*, Habermehl, K. O., Ed., Springer-Verlag, Berlin, 1985, 438.

40. **Walser, K., Prinzen, R., and Brunner, C.**, Preliminary results on the determination of the cell content of milk through ATP measurement by means of bioluminescence methods, *Berl. Muench. Tieraerztl. Wochenschr.*, 91, 373, 1978.

41. **Cullen, G. A.**, Cell counts throughout lactation, *Vet. Rec.*, 83, 125, 1968.

42. **Blackburn, P. S.**, The variation in cell count of cow's milk throughout lactation and from one lactation to the next, *J. Dairy Res.*, 33, 193, 1966.

43. **Ruffo, G., Sangiorgi, F., Moller, F., and Gavazzi, L.**, The influence of the animal's age and the period of lactation on the cell count of milk, *Arch. Vet. It.*, Suppl. 2, 241, 1978.

44. **Blackburn, P. S.**, The cell count of cow's milk and the micro-organisms cultured from the milk, *J. Dairy Res.*, 35, 58, 1968.

45. **Lindstrom, U. B., Kenttamies, H., Arstila, J., and Tuovila, R.**, Usefulness of cell counts in predicting bovine mastitis, *Acta Agric. Scand.*, 31, 199, 1981.

46. **Milne, J. R.**, Observations on the California mastitis test (CMT) reaction. II. Photomicrographic studies of somatic cells and their reaction with surface active agents, *N. Z. J. Dairy Sci. Technol.*, 12, 48, 1977.

47. **Johnston, H. H. and Curtis, G. D. W.**, Detection of bacteria by bioluminescence — problems in removal of non-bacterial ATP, in *Analytical Applications of Bioluminescence and Chemiluminescence, Proceedings 1978*, Schram, E. and Stanley, P., Eds., State Printing and Publishing, Westlake Village, CA, 1979, 446.

48. **Thore, A., Ansehn, S., Lundin, A., and Bergman, S.**, Detection of bacteriuria by luciferase assay of adenosine triphosphate, *J. Clin. Microbiol.*, 1, 1, 1975.

49. **Lundin, A., Hallander, H., Kallner, A., Karnell Lundin, U., and Osterberg, E.**, Bacteriuria detection in an outpatient setting: comparison of several methods including an improved assay of bacterial ATP, in *Rapid Methods and Automation in Microbiology and Immunology*, Habermehl, K. O., Ed., Springer-Verlag, Berlin, 1985, 455.

50. **Lundin, A.**, Extraction and automatic luminometric assay of ATP, ADP and AMP, in *Analytical Applications Bioluminescence and Chemiluminescence*, Kricka, L. J., Stanley, P. E., Thorpe, G. H. G., and Whitehead, T. P., Eds., Academic Press, Orlando, FL, 1984, 491.

51. **Lundin, A., Hasenson, M., Persson, J., and Pousette, A.**, Estimation of biomass in growing cell lines by adenosine triphosphate assay, in *Methods in Enzymology*, Vol. 133, DeLuca, M. A. and McElroy, W. D., Eds., Academic Press, Orlando, FL, 1986, 27.

52. **Linklater, H. A., Galsworthy, P. R., Stewart-DeHaan, P. J., D'Amore, T., Lo, T. C. Y., and Trevithick, J. R.**, The use of guanidinium chloride in the preparation of stable cellular homogenates containing ATP, *Anal. Biochem.*, 148, 44, 1985.

53. **Stanley, P. E.**, Extraction of adenosine triphosphate from microbial and somatic cells, in *Methods in Enzymology*, Vol. 133, DeLuca, M. A. and McElroy, W. D., Eds., Academic Press, Orlando, FL, 1986, 14.

54. **Theron, D. P., Prior, B. A., and Lategan, P. M.**, Effect of minimum growth temperature on the adenosine triphosphate content of bacteria, *Int. J. Food Microbiol.*, 4, 323, 1987.

55. **Lundin, A. and Thore, A.**, Comparison of methods for extraction of bacterial adenine nucleotides determined by firefly assay, *Appl. Microbiol.*, 30, 713, 1975.

56. **Tarkkanen, P., Driesch, R., and Greiling, H.,** A rapid enzymatic micro-method for the determination of intracellular and extra-cellular ATP and its clinical-chemical applications, *Frezenius A. Anal. Chem.,* 209, 180, 1978.

57. **Gheorghiu, M. and Lagranderie, M.,** Mesure rapide de la viabilite due BCG par dosage de l'ATP, *Ann. Microbiol.,* 130B, 147, 1979.

58. **Botha, W. C., Luck, H., and Jooste, P. J.,** The applicability of the adenosine triphosphate method as a rapid bacteriological platform test, *S. Afr. J. Dairy Technol.,* 17, 59, 1985.

59. **Han, S. H., Kim, C. H., Kim, J. B., Shin, H. K., and Lee, S. B.,** Determination of bacterial number in a raw milk by ATP assay monitored by luciferin-luciferase bioluminescence reaction, *Korean J. Anim. Sci.,* 27, 782, 1985.

60. **Zecconi, A., Cecchi, L., and Cugini, F. P.,** Ulteriori osservation sperimentali sulla bioluminescenza applicata alla definizione della qualita igienica del latte, *Arch. Vet. It.* 38, 25, 1987.

61. **Langeveld, L. P. M. and van der Waals, C. B.,** The ATP platform test, the ATP test and the direct microscopic count procedure as methods of estimating the microbial quality of raw milk, *Neth. Milk Dairy J.,* 42, 173, 1988.

62. **Van de Werf, H. and Verstraete, W.,** ATP measurement by bioluminescence: environmental applications, in *Analytical Applications of Bioluminescence and Chemiluminescence,* Kricka, L. J., Stanley, P. E., Thorpe, G. H. G., and Whitehead, T. P., Eds., Academic Press, Orlando, FL, 1984, 33.

63. **Simpson, W. J. and Hammond, J. R. M.,** Cold ATP extractants compatible with constant light signal firefly luciferase reagents, in *Rapid Microbiology Using ATP and Luminescence,* Society of Applied Bacteriology, Tech. Ser. No. 26, Stanley, P., McCarthy, B., and Smither, R., Eds., Blackwell, Oxford, 1989, 45.

64. **Van Crombrugge, J., Waes, G., and Reybroek, W.,** The ATP-F test for estimation of the bacteriological quality of raw milk, *Neth. Milk Dairy J.,* 43, 347, 1989.

65. **Kelbaugh, B. N., Picciolo, G. L., Chappelle, E. W., and Colburn, M. E.,** Automatic bio-sample bacteria detection system, NASA Document TSP 71-10055, National Aeronautics and Space Administration, Washington, D.C., 1971.

66. **Botha, W. C., Luck, H., and Jooste, P. J.,** Determination of bacterial ATP in milk — the influence of adenosine triphosphate-hydrolyzing enzymes from somatic cells and *Pseudomonas fluorescens, J. Food Prot.,* 49, 822, 1986.

67. **Anon.,** Rapid microbial test: bacteria screening kit, Lumac Systems Ag., Basel, 1982.

68. **LaRocco, K. A., Littel, K. J., and Pierson, M. D.,** The bioluminescent ATP assay for determining the microbial quality of foods, in *Foodborne Microorganisms and their Toxins: Developing Methodology,* Pierson, M. D. and Stern, N. J., Eds., Marcel Dekker, New York, 1986, 145.

69. **Quesneau, R.,** La luminescence, *Tech. Lait.,* 974, 49, 1983.

70. **Anderson, M.,** Source and significance of lysosomal enzymes in bovine milk fat globule membrane, *J. Dairy Sci.,* 60, 1217, 1977.

71. **Bogin, E. and Ziv, G.,** Enzymes and minerals in normal and mastitic milk, *Cornell Vet.,* 63, 666, 1973.

72. **Schalm, O. W., Caroll, E. J., Jain, N. C., and Jain, A. H.,** *Bovine Mastitis,* Lea and Febiger, Philadelphia, 1971, chap. 4.

73. **Anon.,** *United Kingdom Dairy Facts and Figures,* 1988 ed., Federation United Kingdom Milk Marketing Boards, Thames Ditton, 1988, 40.

74. **Botha, W. C.,** The Application of the ATP Test for the Assessment of the Bacteriological Quality of Bulk Milk, M.Sc. (Agric.) thesis, University of the Orange Free State, Bloemfontein, 1985.

75. **Chapman, A. D., Fall, L., and Atkinson, D. E.,** Adenylate energy charge in *Escherichia coli* during growth and starvation, *J. Bacteriol.,* 108, 1072, 1971.

76. **Theron, D. P., Prior, B. A., and Lategan, P. M.,** Sensitivity and precision of bioluminescent techniques for enumeration of bacteria in skim milk, *J. Food Prot.,* 49, 8, 1986.

77. **Aledort, L. M. R., Weed, R., and Troup, S. B.,** Ionic effects of firefly bioluminescence assay on RBC ATP, *Anal. Biochem.,* 17, 268, 1966.

78. **Denbury, J. L. and McElroy, W. D.,** Anion inhibition of firefly luciferase, *Arch. Biochem. Biophys.,* 141, 668, 1970.

79. **Lundin, A. and Thore, A.,** Analytical information obtainable by evaluation of the time course of firefly bioluminescence, *Anal. Biochem.,* 66, 47, 1975.

80. **Thore, A.,** Technical aspects of the bioluminescent firefly luciferase assay of ATP, *Sci. Tools,* 26, 30, 1979.

81. **Stanley, P.,** Rapid measurements of bacteria by ATP assay, *Lab. Equip. Dig.,* 20(2), 1982.

82. **Eriksen, B. and Olsen, O.,** Rapid assessment of the microbial status of bulk milk and raw meat with the new instrument: bactofoss, in *ATP Luminescence: Rapid Methods in Microbiology,* Society of Applied Bacteriology, Tech. Ser. No. 26, Stanley, P., McCarthy, B., and Smither, R., Eds., Blackwell, Oxford, 1989, 175.

83. **Jago, P. H., Simpson, W. J., Denyer, S. P., Evans, A. W., Griffiths, M. W., Hammond, J. R. M., Ingram, T. P., Lacey, R. F., Macey, N. W., McCarthy, B. J., Salusbury, T. T., Senior, P. S., Sidorowicz, S., Smither, R., Stanfield, G., and Stanley, P. E.,** An evaluation of the performance of ten commercial luminometers, *J. Biolumin. Chemolumin.,* 3, 131, 1989.

84. **Webster, J., Chang, J. C., Howard, J. L., and Leach, F. R.,** Some characteristics of commercially available firefly luciferase preparations, *J. Appl. Biochem.,* 1, 471, 1979.

85. **Lundin, A.,** Analytical applications of bioluminescence: the firefly system, in *Clinical and Biochemical Luminescence,* Kricka, L. J. and Carter, T. J. N., Eds., Marcel Dekker, New York, 1982, 43.

86. **Bossuyt, R.,** Usefulness of an ATP assay technique in evaluating the bacteriological quality of raw milk, *Kiel. Milchwirtsch. Forschungsber.,* 34, 129, 1982.

87. **Britz, T. J., Bezuidenhout, J. J., Dreyer, J. M., and Steyn, P. L.,** Use of adenosine triphosphate as an indicator of the microbial counts in milk, *S. Afr. J. Dairy Technol.,* 12, 89, 1980.

88. **Labots, H. and Stekelenburg, F. K.,** ATP-bioluminescence: a rapid method for the estimation of microbial contamination of meat and meat products, in 31st European Meet. Meat Res. Workers, Sofia, 1985, 401.

89. **Pernelle, M.,** Utilisations de la bioluminescence en industrie laitiere, *Tech. Lait.,* 980, 33, 1983.

90. **Anon.,** $N_{66}TM$ *Posidyne® Filter Guide,* Pall Process Filtration Ltd, Portsmouth, 1984.

91. **Kroll, R. G.,** Electropositively charged filters for the concentration of bacteria from foods, *Food Microbiol.,* 2, 183, 1985.

92. **Griffiths, M. W.,** Electropositive filters for dairy products, pres. Symp. Separation Pre-treatment Tech. Rapid Methods, London, January 6, 1988.

93. **Olsen, O. and Eriksen, B.,** A newly developed instrument for assessment of the microbial status of raw meat, in Proc. 5th Int. Symp. Rapid Methods Automation Microbiol. Immunol., Florence, 1987.

94. **Wood, J. M. and Gibbs, P. A.,** New developments in the rapid estimation of microbial populations in foods, in *Developments in Food Microbiology — 1,* Davies, R., Ed., Applied Science, London, 1982, 183.

95. **Daniels, S. L.,** The adsorption of microorganisms onto solid surfaces: a review, *Dev. Ind. Microbiol.,* 13, 211, 1972.

96. **Wood, J. M.,** The interaction of micro-organisms with ion exchange resins, in *Microbial Adhesion to Surfaces,* Berkeley, R. C. W., Lynch, J. M., Melling, J., Rutter, P. R., and Vincent, B., Eds., Ellis Horwood, Chichester, England, 1980, 163.

97. **Patel, P. D. and Wood, J. M.,** The potential of chromatographic techniques for the manipulation of viable micro-organisms, in *Rapid Methods and Automation in Microbiology and Immunology,* Habermehl, K. O., Ed., Springer-Verlag, Berlin, 1985, 438.

98. **DeLatour, C. and Kolm, H. H.,** High-gradient magnetic separation a water treatment alternative, *J. Am. Wat. Wks. Assoc.,* 68, 325, 1976.

99. **MacRae, I. C. and Evans, S. K.,** Factors influencing the adsorption of bacteria to magnetite in water and wastewater, *Water Res.,* 17, 271, 1983.

100. **Sharon, N.,** Use of lectins for separation of cells, in *Cell Separation: Methods and Selected Applications,* Vol. 3, Pretlow, T. G., II. and Pretlow, T. P., Eds., Academic Press, San Diego, 1984, 13.

101. **Pistole, T. G.,** Interaction of bacteria and fungi with lectins and lectin-like substances, *Ann. Rev. Microbiol.,* 35, 85, 1981.

102. **Reeke, G. N., Becker, J. W., Cunningham, B. A., Gunther, G. R., Wang, J. L., and Edelman, G. M.,** Relations between the structure and activities of concanavalin A, *Ann. N.Y. Acad. Sci.,* 234, 369, 1974.

103. **Griffiths, M. W., Phillips, J. D., and Muir, D. D.,** Postpasteurization contamination — the major cause of failure of fresh dairy products, *Hannah Res. 1984,* 77, 1985.

104. **Phillips, J. D. and Griffiths, M. W.,** Pasteurized dairy products: the constants imposed by environmental contamination, in *Food Contamination from Environmental Sources,* Nriagu, J. and Simmons, M. S., Eds., John Wiley & Sons, New York, 1989, chap. 13.

105. **Griffiths, M. W. and Phillips, J. D.,** Modelling the relation between bacterial growth and storage temperature in pasteurized milks of varying hygienic quality, *J. Soc. Dairy Technol.,* 41, 96, 1988.

106. **Phillips, J. D., Griffiths, M. W., and Muir, D. D.,** Accelerated detection of post-heat-treatment contamination in pasteurized double cream, *J. Soc. Dairy Technol.,* 36, 41, 1983.

107. **Bishop, J. R. and White, C. H.,** Assessment of dairy product quality and potential shelf-life — a review, *J. Food Prot.,* 49, 739, 1986.

108. **Waes, G. and Bossuyt, R.,** A rapid method to detect postcontamination in pasteurized milk, *Milchwissenschaft,* 36, 548, 1981.

109. **Langeveld, L. P. M., Cuperus, F., Van Breemen, P., and Dykers, J.,** A rapid method for the detection of postpasteurization contamination in HTST pasteurized milk, *Neth. Milk Dairy J.,* 30, 157, 1976.

110. **Waes, G. and Bossuyt, R. G.,** Usefulness of the benzalkon-crystal violet-ATP method for predicting the keeping quality of pasteurized milk, *J. Food Prot.,* 45, 928, 1982.

111. **Phillips, J. D. and Griffiths, M. W.,** Estimation of Gram-negative bacteria in milk: a comparison of inhibitor systems for preventing Gram-positive bacterial growth, *J. Appl. Bacteriol.,* 60, 491, 1986.

112. **Lawton, W. C. and Nelson, F. E.**, The effect of storage temperatures on the growth of psychrophilic organisms in sterile and laboratory pasteurized skim milks, *J. Dairy Sci.*, 37, 1164, 1954.

113. **Rodrigues, U. M. and Pettipher, G. L.**, Use of the Direct Epifluorescent Filter Technique for predicting the keeping quality of pasteurized milk within 24 hours, *J. Appl. Bacteriol.*, 57, 125, 1984.

114. **Oliveria, J. S. and Parmelee, C. E.**, Rapid enumeration of psychrotrophic bacteria in raw and pasteurized milk, *J. Milk Food Technol.*, 39, 269, 1976.

115. **Griffiths, M. W., Phillips, J. D., and Muir, D. D.**, Rapid plate counting technique for enumeration of psychrotropic bacteria in pasteurized double cream, *J. Soc. Dairy Technol.*, 33, 8, 1980.

116. **Phillips, J. D., Griffiths, M. W., and Muir, D. D.**, Pre-incubation test to rapidly identify post-pasteurization contamination in milk and single cream, *J. Food Prot.*, 47, 391, 1984.

117. **Griffiths, M. W., Phillips, J. D., and Muir, D. D.**, Methods for rapid detection of post-pasteurization contamination in cream, *J. Soc. Dairy Technol.*, 37, 22, 1984.

118. **Griffiths, M. W., Phillips, J. D., and Muir, D. D.**, A rapid method for detecting post-pasteurization contamination in cream, in *Analytical Applications of Bioluminescence and Chemiluminescence*, Kricka, L. J., Stanley, P. E., Thorpe, G. H. G., and Whitehead, T. P., Eds., Academic Press, Orlando, FL, 1984, 61.

119. **Phillips, J. D. and Griffiths, M. W.**, Bioluminescence and impedimetric methods for assessing shelf-life of pasteurized milk and cream, *Food Microbiol.*, 2, 39, 1985.

120. **Griffiths, M. W. and Phillips, J. D.**, The application of the pre-incubation test in commercial dairies, *Austral. J. Dairy Technol.*, 41, 71, 1986.

121. **Griffiths, M. W. and Phillips, J. D.**, EEC standards for pasteurized milk — early prediction of passes, *Dairy Ind. Int.*, 53(6), 17, 1988.

122. **Jarvis, B. and Easter, M. C.**, Rapid methods in the assessment of microbiological quality; experiences and needs, *J. Appl. Bacteriol.*, 63 Suppl., 115S, 1987.

123. **Bossuyt, R. and Waes, G.**, Impedance measurements to detect post-pasteurization contamination of pasteurized milk, *J. Food Prot.*, 46, 622, 1983.

124. **Waes, G., Bossuyt, R., and Mottar, J.**, A rapid method for the detection of non-sterile UHT milk by the determination of the bacterial ATP, *Milchwissenschaft*, 39, 707, 1984.

125. **Dabaji, M.**, Bioluminescence entre dans les Ateliers, *L'Usine Novelle*, 41, 138, 1982.

126. **Cogan, T. M.**, A review of heat resistant lipases and proteinases and the quality of dairy products, *Irish J. Food Sci. Technol.*, 1, 95, 1977.

127. **Law, B. A.**, Reviews of the progress of dairy science: enzymes of psychrotrophic bacteria and their effects on milk and milk products, *J. Dairy Res.*, 46, 573, 1979.

128. **Cousin, M. A.**, Presence and activity of psychrotrophic microorganisms in milk and dairy products. A review, *J. Food Prot.*, 45, 172, 1982.

129. **Fairbairn, D. J. and Law, B. A.**, Proteinases of psychrotrophic bacteria: their production, properties, effects and control, *J. Dairy Res.*, 53, 139, 1986.

130. **Stead, D.**, Microbial lipases: their characteristics, role in food spoilage and industrial uses, *J. Dairy Res.*, 53, 481, 1986.

131. **Griffiths, M. W., Phillips, J. D., and Muir, D. D.**, Thermostability of proteases and lipases from a number of species of psychrotrophic bacteria of dairy origin, *J. Appl. Bacteriol.*, 50, 289, 1981.

132. **Law, B. A., Andrews, A. T., and Sharpe, M. E.**, Gelation of ultra-high-temperature-sterilized milk by proteases from a strain of *Pseudomonas fluorescens* isolated from raw milk, *J. Dairy Res.*, 44, 145, 1977.

133. **McKellar, R. C.**, Development of off-flavors in ultra-high temperature and pasteurized milk as a function of proteolysis, *J. Dairy Sci.*, 64, 2138, 1981.

134. **Baker, S. K.**, The keeping quality of refrigerated, pasteurized milk, *Austral. J. Dairy Technol.*, 38, 124, 1983.

135. **Rollema, H. S., McKellar, R. C., Sorhaug, T., Suhren, G., Zadow, J. G., Law, B. A., Poll, J. K., Stepaniak, L., and Vagias, G.**, Comparison of different methods for the detection of bacterial proteolytic enzymes in milk, *Milchwissenschaft*, 44, 491, 1989.

136. **Baldwin, T. O.**, Bacterial luciferase as a generalized substrate for the assay of proteases, in *Methods in Enzymology*, Vol. 57, DeLuca, M. A., Ed., Academic Press, New York, 1978, 198.

137. **Norton, P. T., Merz, R. L., and Leach, F. R.**, The inactivation of firefly luciferase by proteolytic enzymes — an assay for proteases, *Biochem. Int.*, 3, 457, 1981.

138. **Mitchell, G.E. and Ewings, K. N.**, Quantification of bacterial proteolysis causing gelation in UHT-treated milk, *N. Z. J. Dairy Sci. Technol.*, 20, 65, 1985.

139. **Rowe, M. and Pearce, J.**, Personal communication, 1989.

140. **Mottar, J.**, Heat resistant enzymes in UHT milk and their influence on sensoric changes during uncoded storage, *Milchwissenschaft*, 36, 87, 1981.

141. **Kather, H. and Wieland, E.**, Bioluminescent determination of free fatty acids, *Anal. Biochem.*, 140, 349, 1984.

142. **Ulitzur, S.**, A sensitive bioassay for lipase using bacterial bioluminescence, *Biochim. Biophys. Acta*, 572, 211, 1979.

143. **Ulitzur, S.,** A fatty acid-dependent luminescent bacterial mutant for assaying lipolytic activity, in *Analytical Applications of Bioluminescence and Chemiluminescence, Proceedings 1978,* Schram, E. and Stanley, P., Eds., State Printing and Publishing, Westlake Village, CA, 1979, 135.

144. **Ulitzur, S. and Heller, M.,** Bioluminescent assay for lipase, phospholipase A_2, and phospholipase C, in *Methods in Enzymology,* Vol. 72, Lowenstein, J. M., Ed., Academic Press, New York, 1981, 338.

145. **Ulitzur, S. and Hastings, J. W.,** Myristic acid stimulation of bacterial bioluminescence in "aldehyde" mutants, *Proc. Natl. Acad. Sci. U.S.A.,* 75, 266, 1978.

146. **Cardwell, J. T. and Sasso, Y.,** The relation of adenosine tri-phosphate activity to titratable activity in three dairy cultures, *J. Dairy Sci.,* 68(Suppl. 1), 66, 1985.

147. **Nordlund, J., Merilainen, V., and Anderssen, V.,** Bioluminescens — tillampningsmojligheter inom mjolkhushallningen, *Nord. Mejeriindustri,* 7, 5, 1980.

148. **Mackintosh, R. D.,** Personal communication, 1984.

149. **Quesneau, R., Bigret, M., and Luquet, F. M.,** Bioluminescence assay to detect antibiotics and antiseptics in milk, *Spec. Publ. R. Soc. Chem.,* 49, 253, 1984.

150. **Westhoff, D. C. and Engler, T.,** Detection of penicillin in milk by bioluminescence, *J. Milk Food Technol.,* 38, 537, 1975.

151. **Williams, G. R.,** Use of bioluminescence in the determination of the antibiotic content of milk, *J. Soc. Dairy Technol.,* 37, 40, 1984.

152. **Stewart, G., Smith, T., and Denyer, S.,** Genetic engineering for bioluminescent bacteria, *Food Sci. Technol. Today,* 3(1), 19, 1989.

153. **Lee, J.,** The mechanism of bacterial bioluminescence, in *Chemoluminescence and Bioluminescence,* Burr, J. G., Ed., Marcel Dekker, New York, 1985, 401.

154. **Ulitzur, S. and Kuhn, J.,** Introduction of *lux* genes into bacteria, a new approach for specific determination of bacteria and their antibiotic susceptibility, in *Bioluminescence and Chemiluminescence: New Perspectives,* Schölmerich, J., Andreesen, R., Kapp, A., Ernst, M., and Woods, W. G., Eds., John Wiley & Sons, Chichester, England, 1987, 463.

155. **Miyamato, C., Boylan, M., Graham, A., and Meighen, E.,** Cloning and expression of the genes from the bioluminescent system of marine bacteria, in *Methods in Enzymology,* Vol. 133, DeLuca, M. A. and McElroy, W. D., Eds., Academic Press, Orlando, FL, 1986, 70.

156. **Engebrecht, J. and Silverman, M.,** Techniques for cloning and analyzing bioluminescence genes from marine bacteria, in *Methods in Enzymology,* Vol. 133, DeLuca, M. A. and McElroy, W. D., Eds., Academic Press, Orlando, FL, 1986, 83.

157. **Ulitzur, S.,** Determination of antibiotic activities with the aid of luminous bacteria, in *Methods of Enzymology,* Vol. 133, DeLuca, M. A. and McElroy, W. D., Eds., Academic Press, Orlando, FL, 1986, 275.

158. **McMurdo, I. H. and Whyard, S.,** Suitability of rapid microbiological methods for the hygienic management of spray drier plant, *J. Soc. Dairy Technol.,* 37, 4, 1984.

159. **Setlow, P.,** Germination and outgrowth, in *The Bacterial Spore,* Vol. 2, Hurst, A. and Gould, G. W., Eds., Academic Press, London, 1983, 211.

160. **Griffiths, M. W.,** *Listeria monocytogenes:* its importance in the dairy industry, *J. Sci. Food Agric.,* 47, 133, 1989.

161. **Kricka, L. J.,** Clinical and biochemical applications of luciferases and luciferin, *Anal. Biochem.,* 175, 14, 1988.

162. **Shapton, N.,** Hazard analysis applied to control of pathogens in the dairy industry, *J. Soc. Dairy Technol.,* 41, 62, 1988.

163. **Firstenberg-Eden, R., Van Sise, M. L., Zindulis, J., and Kahn, P.,** Impedimetric estimation of coliforms in dairy products, *J. Food Sci.,* 49, 1449, 1984.

164. **Easter, M. C. and Gibson, D. M.,** Rapid and automated detection of salmonella by electrical measurements, *J. Hyg. Cambridge,* 94, 245, 1985.

165. **Gibson, D. M.,** Some modification to the media for rapid automated detection of salmonellas by conductance measurement, *J. Appl. Bacteriol.,* 63, 299, 1987.

166. **Ogden, I. D. and Cann, D. C.,** A modified conductance medium for the detection of *Salmonella* spp., *J. Appl. Bacteriol.,* 63, 459, 1987.

167. **Phillips, J. D. and Griffiths, M. W.,** An electrical method for detecting *Listeria* spp., *Lett. Appl. Microbiol.,* 9, 129, 1989.

168. **Tsai, T. S., Ramey, N. J., and Everett, L. J.,** Rapid separation and quantitation of mixed microorganisms by filtration and bioluminescence, *Proc. Soc. Exp. Biol. Med.,* 183, 74, 1986.

169. **Fung, D. Y. C., Cox, N. A., Goldschmidt, M. C., and Bailey, J. S.,** Rapid methods and automation: a survey of professional microbiologists, *J. Food Prot.,* 52, 65, 1989.

170. **Griffiths, M. W., McIntyre, L., Sully, M., and Johnson, I.,** Enumeration of bacteria in milk, *Proc. VIth Intl. Symp. Biolum. Chemilum.,* Cambridge, September 1990.

171. **Rowe, M. T., Pearce, J., Crone, L., Sully, M., and Johnson, I.,** Bioluminescence assay for psychrotroph protease, *Proc. VIth Intl. Symp. Biolum. Chemilum.,* Cambridge, September 1990.
172. **Kyriakades, A. L., Costello, S. M., Doyle, G., Easter, M. C., and Johnson, I.** Rapid hygiene monitoring using ATP bioluminescence, *Proc. VIth Intl. Symp. Biolum. Chemilum.,* Cambridge, September 1990.
173. **Stewart, G. S. A. B.,** *In vivo* bioluminescence: new potentials for microbiology, *Lett. Appl. Microbiol.,* 10, 1, 1990.

Chapter 3

RAPID FOOD MICROBIOLOGY: APPLICATION OF BIOLUMINESCENCE IN THE DAIRY AND FOOD INDUSTRY — A REVIEW

Ole Olsen

TABLE OF CONTENTS

I.   Introduction 64

II.   ATP as a Tool for the Enumeration of Microorganisms 64

III.   Analytical Procedures to Quantify and Extract ATP from Microbial
        Cells 65
        A.   The Firefly Luciferase Reagents 66
        B.   The ATP Extraction Reagents 66

IV.   Applications of Bioluminescence in the Food Industry 67
        A.   Milk and Dairy Products 68
        B.   Postcontamination in Pasteurized Milk and Cream 69
        C.   Fresh Meat Products and Finished Products 70

V.   BactoFoss Method on Raw Meat 73
        A.   Meat Samples 73
        B.   Homogenization 73
        C.   BactoFoss Method 74
        D.   Reference Method 74
        E.   Results 75
              1.   Pork 75
              2.   Beef 75
        F.   Discussion 76

VI.   Conclusion 77

References 79

I. INTRODUCTION

Food microbiology is a fascinating science because it is very closely related to practical applications in the food industry. The puzzle is to set up an analytical system which can judge the microbial status of a product. The information is used to accept or reject raw materials and finished products as well as to control the process line. The results, together with a lot of practical experience, enable you to judge whether a product is of good or bad quality.

It is important to remember that microorganisms in a food product live in microenvironments defined on scales of micrometer and millimeter dimensions. The physical and chemical characteristics of these microhabitants are generally quite different from those of the ambient microenvironments or the environments established in the enumeration procedures used in classical food microbiology.

The development of reliable techniques, which can be used to enumerate microorganisms and estimate their growth rates, is still considered to be the most fundamental research objective of experimental food microbiology.

Extensive laboratory studies have been conducted in an attempt to characterize the physiology and biochemistry of bacteria isolated from food products. However, it is unknown whether or not these *in vitro* metabolic potentials are ever realized *in situ* among the naturally occurring microbial populations. Methods and conditions for enumeration of microorganisms of food origin are numerous and especially the use and value of rapid instrumental methods in food microbiology are a matter of serious interest.

This review emphasizes a presentation and discussion of the use of adenosine 5′-triphosphate (ATP) measured by bioluminescence in experimental food microbiology — the BactoFoss method will be dealt with in detail.

The present effort is intended to

1. Offer a brief review of the fundamental physiological principles which provide the motivation for using ATP measurements in experimental food microbiology
2. Outline the principles of analytical procedures used to extract and quantify ATP from the microbial assemblages present in foods
3. Present specific applications and interpretations of the measurement of ATP in naturally occurring populations of microorganisms in foods

II. ATP AS A TOOL FOR THE ENUMERATION OF MICROORGANISMS

The use of ATP as an index of microbial biomass is based on three important assumptions:

1. All living organisms contain ATP.
2. ATP is neither associated with dead cells nor absorbed onto surfaces, colloids, etc.
3. There exists a fairly constant ratio of ATP to biomass/number of cells for all microbial taxa independent of metabolic activity or environmental conditions.

Of these assumptions only the first one is unequivocally acceptable. Assumption number two concerning the obligate association of ATP with living organisms must be modified.[1-3]

Many food products, including raw meat, poultry, and shellfish, contain certain amounts of intrinsic (nonmicrobial) ATP, and fresh meat samples contain highly variable amounts of intrinsic ATP depending upon the cut and age of the meat.[4] The ATP background level in a meat sample may be equivalent to 10^5 to 10^8 cfu/g. The nature of this intrinsic ATP varies from an easily explainable form, e.g., in living somatic cells in raw milk,[5] to more complex forms, e.g., in meat homogenates where interfering nonmicrobial ATP may be present associated with large or fine collodial meat particles[3] or micelles in milk.[6]

The third premise has been reviewed thoroughly.[1,7] The most critical and most widely criticized assumption of the ATP-bacterial enumeration assay is the validity of the application of a single ATP/cfu ratio for converting ATP measurements to estimated cfu. There is an overwhelming amount of literature documenting that the ATP/cfu ratio varies considerably between individual taxa and even within certain species as a function of culture conditions and growth rate. On the other hand, correlations between ATP and cell/biomass measurements could not be drawn if intracellular ATP levels changed radically throughout the growth cycle or as a function of culture conditions.

The turnover time of ATP in growing microorganisms is reported to be 1 s or less[7] and reactions in which ATP is utilized and regenerated in the cell are numerous. There is, however, much evidence that a steady-state balance for intracellular ATP exists to provide sufficient energy for maintenance of respiratory and metabolic enzymes during all phases of growth.[1,7] This is accomplished through precise regulation of the metabolic energy stored in the ATP pool. Therefore, correlating intracellular ATP concentrations with cfu is a relatively simple matter under defined and comparable experimental conditions. Many growth curve studies involving ATP measurements and conventional bacteria cfu determinations illustrate that ATP levels increase parallel with cfu during the logarithmic phase of growth.[2,5,8,9]

There have been numerous studies on the quantization of the steady-state ATP levels of a variety of microorganisms.[1,7,8,10] Even though assay conditions, reagents, and equipment differed in each study, the ATP content in femtogram (fg) ATP/cfu (1 fg = 10^{-15} g) were found to range from approximately 0.1 to 4.0, with an average of approximately 1 fg ATP/cfu for bacteria. Different suboptimal conditions of the microorganisms involving injury, lag, or declining phases may change the ratio ATP/cfu outside the above-mentioned range.

Although the accuracy of the ATP-enumeration technique of bacteria has been questioned and numerous difficulties in its application have been cited, it is still, as discussed below, a convenient and reliable method for enumerating bacteria.

III. ANALYTICAL PROCEDURES TO QUANTIFY AND EXTRACT ATP FROM MICROBIAL CELLS

The firefly (*Photuris pyralis*) bioluminescence system has been described thoroughly in detail by DeLuca and McElroy[11] and by other authors in this book and will not be dealt with in this paper.

The use of bioluminescence to determine ATP is by no means a new technique. But, it was not until the early 1980s that the greater utilization of bioluminescence went from the research laboratories into application and control laboratories and the technique began to be a routine analytical tool. There are at least two important factors which are responsible for this progress:

1. The availability of commercially prepared reagent kits
2. The commercial manufacture of more suitable analytical equipment meant for use in routine laboratories

The two advantages stressed above have become more and more refined, with suppliers developing application notes and reagent kits linked and optimized to a particular piece of equipment.

This process has developed in steps from very flexible luminometers where reagent and sample manipulations are fully manual, to semi-automatic luminometers with built in reagent handling, to the BactoFoss method where both sample treatments and reagent handlings are fully automatic and the method is dedicated to well-defined customer groups in the food industry.

A. THE FIREFLY LUCIFERASE REAGENTS

In the early days of firefly bioluminescence the users made their own preparations of crude extracts from firefly tails or lanterns. The enzyme stability in these crude solutions did vary considerably.[11] Later, commercial preparations of these crude extracts were also available. In the late 1960s an abbreviated purification scheme was developed that removed ATP (and thus reduced the background) and interfering enzymes.[11] In the middle to late 1970s purified and even crystalline firefly luciferase preparations began to be commercially available from several suppliers and the technique began to be accessible outside research laboratories.[11]

From the late 1970s up until today, the field of firefly bioluminescence has developed rapidly with respect to reagent formulation.[12] The manufacturers of reagents, to a great extent, identified a potential for the use of the firefly bioluminescence technique in quality and process control laboratories where the staff is relatively nonskilled.

These laboratories require methods and reagents which are easy to operate. The reagents need to be almost immediately ready to use, have a long storage time on stock, and be stable not less than a working day after reconstitution. Most manufacturers have chosen the kit format with attuned reagents, buffers, bottles, etc. optimized for one working day. In this way a quality product can be guaranteed, and it is easy to operate by relatively nonskilled people. Each manufacturer's kit has its specific differences in composition and in the philosophy of reagent formulation. The differences are significant for certain experimental systems, e.g., certain applications, or attuned to a specific luminometer. Typically, only the manufacturer knows the exact formulation.

However, due to the properties of firefly luciferase, certain additives are required to stabilize the enzyme and make it usable over a period of time. Other additives are required to optimize the reaction conditions and reaction kinetics.[12] The firefly luciferase is insoluble in water and therefore requires some salts to be solubilized. The pH optimum of firefly luciferase is 7.8, most often using organic buffers (Tris, HEPES, MOPS, Tricine, glycylglycine). The kits contain Bovine Serum Albumin and a sulfhydryl compound, e.g., dithiothreitol, to stabilize the activity of the firefly luciferase.

D-Luciferine, together with ATP, are substrates in the enzymatic reactions with added magnesium ions as co-factor. EDTA is often added in order to prevent an undesirable inhibition by other metal ions, which can also shift the wavelength of the emitted light.

The above-mentioned reagents can be optimized in many ways. The reagents must also be optimized depending upon the procedure used to extract ATP from the cells. Optimization procedures must consider the nature of the analytical equipment, the reaction kinetics, and the robustness of the assay.

B. THE ATP EXTRACTION REAGENTS

Before being accessible to enzymatic analysis the intracellular ATP has to be extracted from the bacterial cell. This is usually achieved by addition of a chemical substance, i.e., an extractant which penetrates the cell walls and cell membranes of the bacteria.[1,13-15] A reliable extractant fulfills three requirements:

1.   It releases the entire intracellular pool of ATP to be assayed.
2.   It gives a complete and rapid inactivation of all enzyme systems in the extracted cells that may affect the ATP content from extraction through analysis of the extracts.
3.   It may not inhibit or affect the kinetics of the firefly reaction in a manner which makes standardization a problem, or extract undue materials which quench or inhibit the reaction.

Numerous methods are available for the extraction of ATP from microbial cells, and there have been a few critical studies and reviews.[1,13-15] The extractants can be grouped into (1) boiling aqueous buffers, (2) diluted strong inorganic and organic acids, (3) organic compounds, and (4) surfactants and mixtures thereof.

No single procedure has emerged as the most efficient in all situations. The boiling buffers and the strong acids are popular because of simplicity. They are quite satisfactory if very high sensitivity is not required. The relative high concentration of a buffer or an acid has to be reduced and neutralized before the firefly reaction is initiated. This normally involves a dilution of the sample, which is a disadvantage when high sensitivity is required.

The use of surfactants and combinations of several surfactants has the possibility of extraction of the bacterial ATP in very small volumes. Then the firefly reaction can be carried out without any dilution and achieve the maximal sensitivity. Many detergents influence the kinetics of the firefly reaction.[16]

Combinations of disinfectants and surfactants can be optimized in several ways to achieve an efficient extraction of ATP from the microbial cell, as well as a controlled and reproducible kinetics of the firefly reaction without dilution of the extract. However, it is important not to overload the extractant with too much sample. It is important to check the sampling/ extraction procedure to ascertain that the process itself does not influence the measured result.

This latter-mentioned philosophy using disinfectant-surfactant combinations is included in most of the newer procedures being supplied as kits by several manufacturers. These kits are often optimized only for a limited number of applications specified by the manufacturers. The reagents in the kit are attuned to the applications, and often also to specified analytical equipment, which performs an attuned process optimizing the kinetics of the firefly reaction. It may not be advisable to change procedures and/or suppliers of reagents because of the customized nature of the reagents.

IV. APPLICATIONS OF BIOLUMINESCENCE IN THE FOOD INDUSTRY

Before the firefly reaction can be used to estimate the microbial contamination levels in food products, the specimen of interest normally has to go through a preparation procedure.[3] In relation to the microbial contamination level, the sample types can be categorized into three different groups:

1. Those containing microbial cells only, e.g., starter cultures, process water, fermentation liquors
2. Those containing both microbial and somatic cells, e.g., most raw food products
3. Those containing low levels of bacteria, e.g., pasteurized or sterilized products

All these categories of samples are analyzed routinely in the food industry, and research has been done to apply the firefly reaction to all of them for estimating the microbial contamination level. General remarks and problems to consider in dealing with sample pretreatment will be dealt with below.

Samples containing microbial cells only are relatively simple in their application to the firefly reaction. The cell number is often high or the sample matrix is simple which makes it easy to concentrate cells by mild centrifugation or filtration.[1] The following ATP assay is straightforward, due to the high number of cells present.

Samples containing a mixture of microbial and somatic cells are more difficult because it is necessary to differentiate between microbial and nonmicrobial ATP. High levels of nonmicrobial ATP are particularly a problem in raw and fresh foods. There are two possible approaches to this. One is to separate the microorganisms from the food before extraction and determination of their ATP and the other is to selectively extract and destroy or remove the nonmicrobial ATP before carrying out the ATP assay. Both of these philosophies have been attempted in different procedures.

The selective extraction and destruction of nonmicrobial ATP is based on the fact that somatic cells in the sample are less resistant to physical stress, e.g., pressure, filtration, osmotic shock, and chemical treatment, e.g., treatment with surfactants, than the microbial cells.[9,17,18]

The nonmicrobial ATP can be destroyed by ATP-degrading enzymes (apyrases) after release from the somatic cells, or the bacterial cells can be removed from the somatic ATP by centrifugation, filtration, or cation-exchanging techniques. These later-mentioned physical separation methods can also be used to concentrate bacteria from food products which have a lower number of viable organisms than the detection level of the ATP assay itself and thereby, expand the dynamic range of the technique toward a lower detection level.

A. MILK AND DAIRY PRODUCTS

Enumeration of bacteria in raw or processed milk is of public health and economic importance. For some time the dairy industry has required a bacterial counting procedure to monitor tankers' milk arriving at processing sites or central collection facilities before intra- or inter-country shipment. The test required should take less than 5 to 10 min to provide results before the tanker unloads.

With the introduction of refrigerated milk storage, a change in bacterial content has occurred and the rapid dye reduction methods previously used to indicate numbers of bacteria are now useless.[19]

The ATP bioluminescence technique has the potential to be a solution as a platform test for the inspection of incoming raw milk to a dairy, control of milk in the milk storage silos, and environmental and process control in the processing plant. The technique is sufficiently fast, it has the potential for sufficient sensitivity, and the number of samples at a plant per day is limited. Therefore, the relatively high cost price per sample can be justified. The detection of bacteria in milk by this method is complicated by the presence of free ATP associated with the colloidal calcium phosphate-citrate complex of the casein micelle,[6] and also by the presence of somatic cell ATP, both of which must be removed before meaningful estimates of bacterial ATP can be obtained.

The hygienic quality of raw milk in most countries in Western Europe and North America has increased very rapidly in recent years. In 1989 only a few countries had numbers above 10^5 cfu/ml raw milk, and within the Scandinavian countries and the U.K. most farmers collect samples which are within the range 10^4 to 3×10^4 cfu/ml for raw milk. The somatic cells in milk are approximately ten times larger in volume than the bacteria and a number of 2×10^5 somatic cells per ml is realistic in raw milk with a good bacteriological quality. Therefore, it is very important that the sample pretreatment procedure to remove somatic cell ATP is a well-controlled and reliable process.

The first method with promising results applied to milk was developed by Bossuyt in 1981.[20] It was a 20 min test. The correlation coefficient between bacterial ATP content and the total colony count was 0.93 both for 48 milk tankers and 209 farm milk samples. The correlation coefficient was less clear in milk samples having a total colony count of 10^5 cfu/ml or less due to inefficient removal of ATP from somatic cells. This procedure was modified afterward to provide a tool for rapid quality assessment at milk reception as the so-called "ATP platform test".[21] This test could be performed within 10 min with a correlation coefficient of 0.83, but suffered the disadvantage of failing to correlate with colony counts below 10^6 cfu/ml milk.

The sample pretreatment was improved by optimizing the procedure and including an apyrase (an ATPase) to destroy the released somatic cell ATP.[22] Although the pretreatment resulted in 96% reduction in the ATP content of raw milk, the remaining nonbacterial ATP was still considerably more than normally found in the bacterial component of the raw milk.

The above-cited techniques describe the ATP assay on whole raw milk samples. This concept can be used for bacteria loads of 10^5 cfu/ml milk and milk samples at contamination levels above this number. There are three major difficulties in the determination of bacterial cell numbers in milk lower than 10^5 cfu/ml:

1.   Quenching of the light production from ATP
2.   The presence of nonbacterial ATP yielding a high background value, mainly from somatic cell ATP
3.   The number of bacteria required for an accurate ATP measurement

Recently, four papers have been published which include a filtration step to overcome these problems.[5,9,23,24] The bacteria are isolated and concentrated on a membrane filter free from nonmicrobial ATP, somatic cells, and quenching materials. Consequently, the sensitivity is improved down to 10,000 to 20,000 cfu/ml milk.

The BactoFoss instrument (Figure 1) is a fully automated instrument using the ATP-bioluminescence technique with a built-in filtration sample preparation unit. The operation procedures in the BactoFoss are schematically illustrated in Figure 2 and Figure 3. The instrument automatically takes a fixed volume from the presented milk sample and deposits it in a temperature-controlled funnel where several filtrations and sample treatments take place to remove somatic cell ATP. The microorganisms are finally isolated on a filter paper, transported internally, and positioned in an extraction chamber where an extraction liquid is added to release the ATP from the microbes. The amount of ATP is determined by measuring the luminescence after addition of luciferin and luciferase. Reagents and consumables attuned to the instrument are supplied by Foss Electric.

The validation of the BactoFoss method has been documented by growth curve studies where ATP measurements and conventional bacterial cfu determinations showed a strong correlation.[9]

A linear relationship between log (light counts) and log (cfu) for 185 samples of raw milk from individual tankers is shown in Figure 4. Ten of the samples have been stored for 24 h at 6°C before they were analyzed so as to expand the dynamic range from 10^4 to 10^6 up to 10^4 to 10^8 cfu/ml. The correlation coefficient was 0.92, the slope of the regression line was 1.17, and the residual standard deviation (s_{yx}) for the log plate count was 0.28 calculated via the regression line from the log light units.

The instrument was placed at a dairy in Germany to investigate how variation both in the composition of the milk and the flora would affect the results. Individual samples (119 of raw milk from incoming tankers) were analyzed (Figure 5). The correlation with plate count was good again, with a correlation coefficient of 0.88 and a residual standard deviation (s_{yx}) for the log plate count of 0.21 calculated via the regression line from the log light units.

The results given in Figure 6 are a composite presentation of all the milk data (a total of 304 samples) to indicate the overall performance of the instrument. They show a correlation coefficient of 0.92. The residual standard deviation (s_{yx}) for the log plate count was 0.27 calculated via the regression line from log light units. The slope of the regression line was 1.17. The BactoFoss method has a dynamic range of 10^4 to 10^8 cfu/ml of raw milk.

B. POSTCONTAMINATION IN PASTEURIZED MILK AND CREAM

It is well documented that the main factors governing the shelf-life of pasteurized milk products are the growth of psychrotrophic, Gram negative bacteria introduced as a result of post-heat-treatment contamination.[25] Consequently, the dairy industry is greatly interested in a rapid and simple method to detect postcontamination in pasteurized milk.

FIGURE 1. The BactoFoss instrument. It is a fully automated instrument performing the ATP assay with a built-in preparation unit.

At Hannah Research Institute, Scotland, a pre-incubation test has been developed which can be used in commercial dairies. The test is suitable for predicting shelf-life of freshly pasteurized milk and cream within 26 h of production.[26,27]

The pre-incubation period allows a selective growth of psychrotrophs at a temperature of 21°C in 25 h with incorporation of an inhibitory system (crystal violet-penicillin-nisin) which prevents Gram positive bacterial growth. The bacteria that are growing under these pre-incubation conditions can be estimated using the ATP-bioluminescence system. A similar test for detection of postcontamination and predicting the keeping quality of pasteurized milk has been described using a stronger selective system (Benzalkon-crystal violet) at a higher temperature 30°C after 24 h.[28,29]

These rapid tests are important from a finished product's point of view but certainly are also valuable when used as process control, monitoring the efficiency of the plant's cleaning operations.

C. FRESH MEAT PRODUCTS AND FINISHED PRODUCTS

In the food industry microbial examination of raw materials, plant surfaces, and finished products is directed toward the maintenance of good manufacturing practice. The results of such examinations are reported to the process controllers who take corrective actions if necessary. The process controllers need results as early as possible. Therefore, raw fresh

Operating Procedure

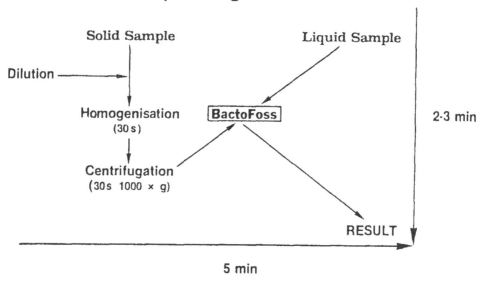

FIGURE 2.   Operating procedures in the BactoFoss method.

foods were among the first products to be studied by food microbiologists when evaluating the ATP-bioluminescence technique as a tool for rapid monitoring of the microbial status of foods.

However, early attempts to estimate the microbial content of foods by determination of microbial ATP were less successful because of the interference from high amounts of nonmicrobial ATP in fresh foods. It was concluded by early researchers that some means of differentiating between microbial and nonmicrobial ATP were necessary for these types of food samples.[30] One approach was to separate the microorganisms before extraction and determination of their ATP. The other approach was to selectively extract and destroy the nonmicrobial ATP first.[31]

The approach where somatic cells in raw meat are selectively extracted for ATP by surfactants like Triton® X-100 or different commercial proprietary reagents, followed by enzymatic destruction of this ATP, before measurement of the bacterial ATP, has been described in detail elsewhere.[8,32,33] This experimental procedure used on raw meat samples produced reasonable correlation coefficients to standard plate counts and seemed applicable without restrictions other than the lower limit of the reliable result of approximately 5×10^6 microorganisms per gram meat sample. About 100 samples per day can be assayed using a manual technique involving one person. The result of one analysis can be available within 1 h at a reasonable price per test.

Different separation procedures to remove microorganisms from food homogenates have been studied on a variety of food samples. In clear liquids like carbonated beverages, artificially contaminated with yeast, separation is easily achieved by membrane filtration to collect microorganisms before extraction of their ATP.[34] Studies on the separation of microorganisms from more complex products normally also include centrifugation, or filtration of the homogenates through glass wool or coarse filters to remove large meat particles, before the microorganisms are collected on a bacteriological membrane filter.[35] Stannard and Wood used a three-stage separation technique.[31] A short centrifugation was used to remove large food debris. A cation-exchange resin was used to absorb small food particles,

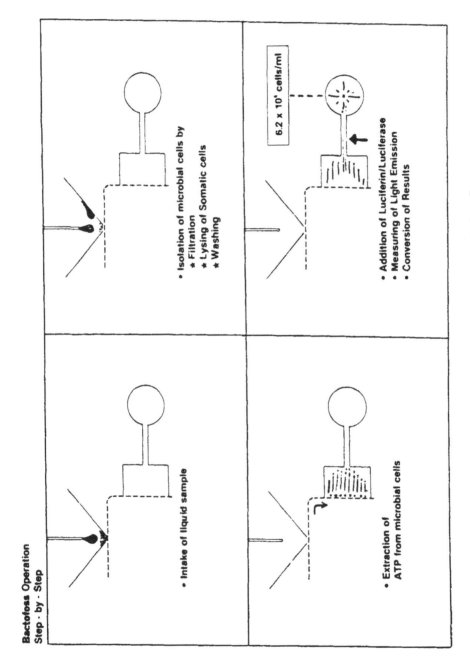

FIGURE 3.   Schematic illustration of sample treatment in the BactoFoss.

Log (CFU/ml)

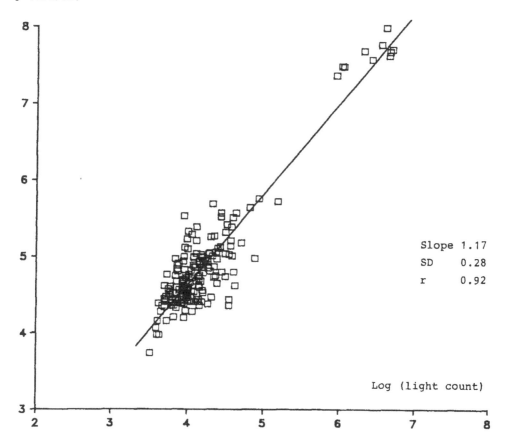

FIGURE 4. Raw milk from incoming tankers to a Danish dairy, 185 individual samples, bioluminescence vs. TVC. The ten most contaminated samples have been incubated 24 h at 6°C. X-axis: log (light count). Y-axis: log (cfu/ml).

and the bacteria remaining in the supernatant effluent were concentrated onto a membrane filter before bacterial ATP extraction. These techniques have been used to analyze many different products, e.g., beef, pork, lamb, and chicken. Tests showed a very high linear correlation to the standard plate count. The lower limit of reliable results was approximately 10^5 microorganisms per gram meat sample.

V. BACTOFOSS METHOD ON RAW MEAT

The BactoFoss method combines the two approaches mentioned above in order to remove the ''noise'' coming from the sample itself. The performance of the BactoFoss method on raw pork and beef will be dealt with in more detail below.

A. MEAT SAMPLES

The trial on raw pork was performed at a large Danish meat processing plant. The size of the meat lombs ranged from a few grams up to 200 to 400 g. The beef meat used for the trial was obtained locally from retail dealers. The meat lombs were typically 150 to 500 g in size. Samples were taken from the surface and the inner part of the meat lombs.

B. HOMOGENIZATION

Ten grams of meat were diluted with 90 g of dilution liquid (8.5 g of NaCl and 1.0 g

Log (CFU/ml)

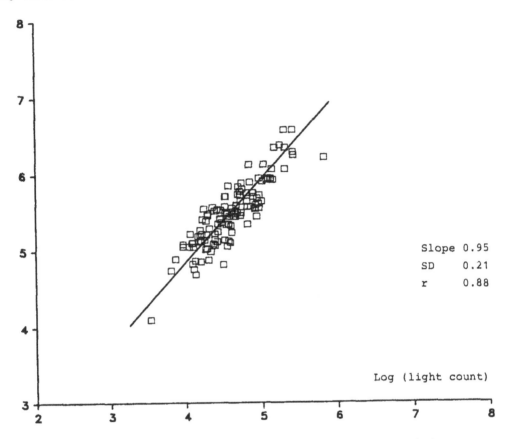

FIGURE 5.   Raw milk from incoming tankers to a German dairy, 119 individual samples bioluminescence vs. TVC. X-axis: log (light count). Y-axis: log (cfu/ml).

of peptone in 1 l of distilled water) and homogenized for 30 s in a Stomacher® (Colworth 400). This suspension was used for the BactoFoss and the reference method, respectively.

C. BACTOFOSS METHOD

A small portion of the meat suspension was centrifuged for 30 s at $350 \times g$, in order to remove debris and coarse meat particles. The sample was placed in the BactoFoss (Foss Electric). The instrument automatically takes out the necessary sample volume and performs the measurement. The different steps in a BactoFoss measurement are illustrated in Figure 2. After 1.5 min the result appeared on the display and was printed out. The instrument automatically performed a rinsing cycle (1.5 min) and was ready for a new sample. A measurement consisted of the following operations: intake of sample; filtration; lysing of somatic cells; washing; extraction of microbial ATP; measuring of light; presentation of result; and rinsing of the instrument.

D. REFERENCE METHOD

The total viable count (TVC) was determined by the spiral plating technique (Spiral System) on Bacto Plate Count Agar (PCA, Difco). The necessary dilutions were made with the same dilution liquid used for the homogenization. The plates were incubated at 21°C for 4 days and counted manually.

Log (CFU/ml)

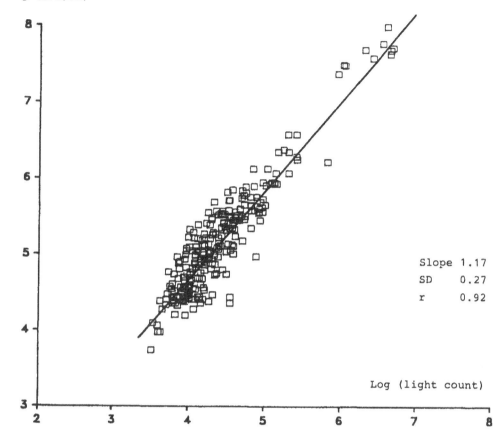

Slope 1.17
SD   0.27
r   0.92

Log (light count)

FIGURE 6. Raw milk from incoming tankers, a composite presentation of 304 individual samples, biolumi-nescence vs. TVC. The ten most contaminated samples have been incubated 24 h at 6°C. X-axis: log (light count). Y-axis: log (cfu/ml).

E. RESULTS

1. Pork

The new method (BactoFoss) was tested at a large Danish meat processing plant. Some 70 samples of pork were examined on the BactoFoss and PCA. The level of contamination ranged from 3000 to 50,000,000 cfu/g. The correlation between the two methods is shown in Figure 7.

The line drawn in Figure 7 was the calibration line which was used by the instrument to estimate the TVC according to the amount of light measured.

The residual standard deviation (s_{yx}) was 0.39 log cycles. If the calibration mode was used and the BactoFoss estimates were correlated with the TVC, the data shown in Figure 8 appeared. A linear regression in the range of 10^5 to 5×10^7 cfu/g gave a correlation coefficient of 0.93 and a residual standard deviation (s_{yx}) of 0.23 log cycles.

2. Beef

The examined beef had a level of contamination from 7×10^2 to 10^9 cfu/g. Figure 9 illustrated the correlation between the raw data from the BactoFoss and the TVC. The residual standard deviation (s_{yx}) was 0.47 log cycles. If a linear regression on the calibrated data was performed it gave a correlation coefficient of 0.94 (Figure 10).

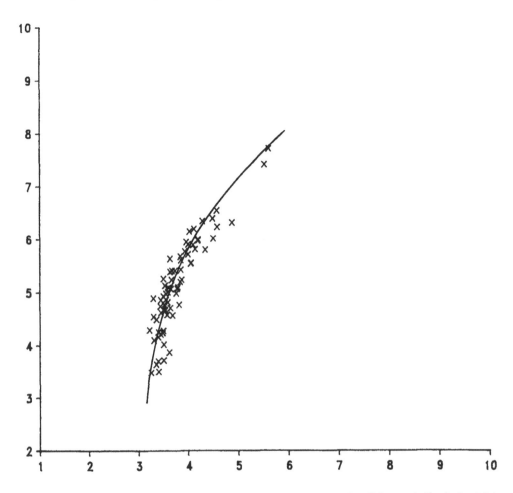

FIGURE 7.   Pork, 70 individual samples, bioluminescence vs. TVC. X-axis: log (light count). Y-axis: log (cfu/ml).

F. DISCUSSION

The correlation between the BactoFoss method and the traditional method (plate count agar) was linear down to 10^5 cfu/g, Figures 7 and 9. Below 10^5 cfu/g the relationship was no longer linear. This was probably due to residual somatic cell ATP which had not been removed prior to the measurement of the microbial ATP.

The interference from somatic cell ATP has been shown by several other investigations.[8,32,33] To take this deviation from the linear correlation into account, the BactoFoss uses a curve fit for its calibration programs.

The pork meat with TVC below 10^4 cfu/g was estimated by the BactoFoss to a level which was approximately ten times higher than the TVC from PCA. This was probably due to the fact that somatic ATP was being measured as microbial ATP.

The data obtained on beef (Figures 9 and 10) resembled the pork data. The correlation was linear down to 10^5 cfu/g and deviated from a straight line at lower levels.

The general results obtained on meat with the BactoFoss method, showed that the procedure has great potential as a rapid and reliable microbiological method. The results obtained on various kinds of beef and pork have demonstrated the possibility of getting a count within 5 min with this method. The BactoFoss is capable of giving a reliable result if the samples have a level of contamination above 10^5 cfu/g. For samples with TVC below 10^5 cfu/g the instrument can confirm this.

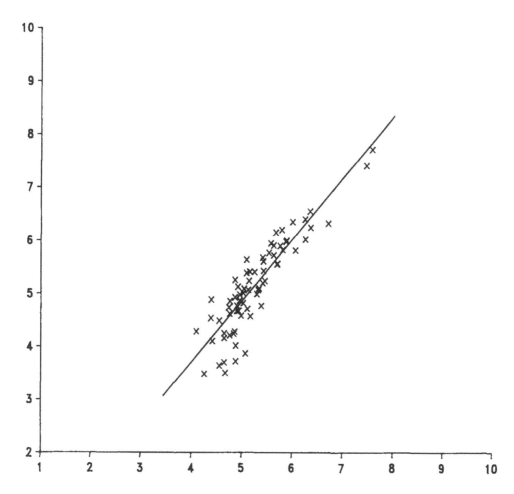

FIGURE 8.   Correlation between BactoFoss (pork calibration) estimates and TVC for 70 individual samples. X-axis: log (estimate). Y-axis: log (cfu/g).

Investigations on raw bulk milk have shown a linear correlation for the range 10^4 to 10^8 cfu/ml with an estimated standard deviation of 0.27 log cycles (Figure 6). This indicates that the BactoFoss is a versatile instrument for the industry. Rapid microbiology close to the processing line has become a realistic possibility with the aid of this new instrument. The simplicity of operating a BactoFoss will certainly be appreciated by the user.

VI. CONCLUSION

The firefly ATP bioluminescent assay has several distinct advantages when compared to other rapid microbiological detection/enumeration procedures. The methodology is rapid and sensitive, and automated equipment with standardized reagent kits is now available to perform the ATP assay. The test can be operated by people without microbiological experience and used as a tool in process control, where quick results are essential, e.g., rejection of raw materials on the platform or diverting raw materials to the correct process line. The technique has the potential to be used as a tool in the industry to assess and improve the hygienic quality of food products.

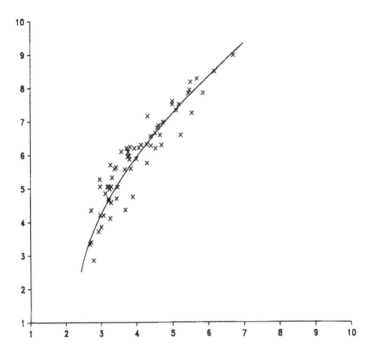

FIGURE 9.   Beef meat, 65 individual samples, bioluminescence vs. TVC. X-axis: log (light unit). Y-axis: log (cfu/ml).

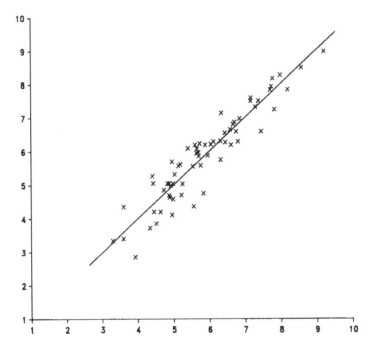

FIGURE 10.   Correlation between BactoFoss (beef meat calibration) estimates and TVC for 65 individual samples of beef meat. X-axis: log (estimate). Y-axis: log (cfu/ml).

REFERENCES

1. **Karl, D. M.,** Cellular nucleotide measurements and application in microbial ecology, *Microbiol. Rev.,* 44, 739, 1980.
2. **LaRocco, K. A., Littel, K. J., and Pierson, M. D.,** The bioluminescent ATP assay for determining the microbial quality of foods, in *Food Born Micro-organisms and Their Toxins,* Pierson, M. D. and Stern, N. J., Eds., Marcel Dekker, New York, 1986, 145.
3. **Stannard, C. J. and Gibbs, P. A.,** Rapid microbiology: application of bioluminescence in food industry — a review, *J. Biolumin. Chemilumin.,* 1, 3, 1986.
4. **Hamm, R.,** Postmortem changes in muscle with regard to processing of hot-boned beef, *Food Technol.,* 36, 105, 1982.
5. **Eriksen, B. and Olsen, O.,** Rapid assessment of the microbial status of bulk milk and raw meat with the new instrument: BactoFoss, in *ATP Luminescence: Rapid Methods in Microbiology, Soc. Appl. Bacteriol., Tech. Ser., Vol. 26,* Stanley, P. E., McCarthy, B. J., and Smither, R., Eds. 1990, 175.
6. **Richardson, T., McGann, T. C. A., and Kearney, R. D.,** Level and location of adenosine 5'-triphosphate in bovine milk, *J. Dairy Res.,* 47, 91, 1980.
7. **Chapman, A. G. and Atkinson, D. E.,** Adenine nucleotide concentrations and turnover rates. Their correlation with biological activity in bacteria and yeast, *Adv. Microb. Physiol.,* 15, 253, 1977.
8. **Bülte, M. and Reuter, G.,** The bioluminescence technique as a rapid method for the determination of the microflora of meat, *Int. J. Food Microbiol.,* 2, 371, 1985.
9. **Olsen, O. and Eriksen, B.,** Introduction of BactoFoss, a newly developed instrument for assessment of the microbial status of raw meat, in *Rapid Methods and Automation in Microbiology and Immunology,* Balows, A., Tilton, R. C., and Turano, A., Eds., Brixia Academic Press, Brescia, Italy, 1989, 746.
10. **Theron, D. P., Prior, B. A., and Lategan, P. M.,** Effect of temperature and media on adenosine triphosphate cell content in *Enterobacter aerogenes, J. Food Prot.,* 46, 196, 1983.
11. **DeLuca, M. and McElroy, W. D.,** Purification and properties of firefly luciferase, *Methods Enzymol.* 157, 3, 1978.
12. **Leach, F. and Webster, J. J.,** Commercially available firefly reagents, *Methods Enzymol.,* 133 B, 51, 1986.
13. **Lundin, A.,** Extraction and automatic luminometric assay of ATP, ADP, and AMP, in *Analytical Applications of Bioluminescence and Chemiluminescence,* Kricka, L. J., Stanley, P. E., Thorpe, C. H. G., and Whitehead, T. P., Eds., Academic Press, Orlando, FL, 1984 491.
14. **Lundin, A. and Thore, A.,** Comparison of methods for extraction of bacterial adenine nucleotides determined by firefly assay, *Appl. Microbiol.,* 30, 713, 1975.
15. **Stanley, P. E.,** Extraction of adenosine triphosphate from microbial and somatic cells, *Methods Enzymol.,* 133 B, 14, 1986.
16. **Kricka, L. J. and DeLuca, M.,** Effect of solvents on the catalytic activity of firefly luciferase, *Arch. Biochem. Biophys.,* 217, 674, 1982.
17. **Stannard, C. J.,** ATP assay as a rapid method for estimating microbial growth in foods, in *Analytical Applications of Bioluminescence and Chemiluminescence,* Kricka, L. J., Stanley, P. E., Thorpe, G. H. J., and Whitehead, T. P., Eds., Academic Press, Orlando, FL, 1984, 53.
18. **Manger, P. A.,** Preliminary evaluation of the Lumac Biocounter M2010 system, measuring bacterial ATP in food, using a bioluminescent technique, in *Analytical Applications of Bioluminescence and Chemiluminescence,* Kricka, L. J., Stanley, P. E., Thorpe, G. H. J., and Whitehead, T. P., Eds., Academic Press, Orlando, FL, 1984, 57.
19. **O'Toole, D. K.,** Methods for the direct assessment of the bacterial content of milk, *J. Appl. Bacteriol.,* 55, 187, 1983.
20. **Bossuyt, R.,** Determination of bacteriological quality of raw milk by an ATP assay technique, *Milchwissenschaft,* 36, 257, 1981.
21. **Bossuyt, R.,** A 5-minute ATP platform test for judging the bacteriological quality of raw milk, *Neth. Milk Dairy J.,* 36, 355, 1982.
22. **Theron, D. P., Prior, B. A., and Lategan, P. M.,** Determination of bacterial levels in raw milk: selectivity of non-bacterial ATP hydrolysis, *J. Food Prot.,* 49, 4, 1986.
23. **Webster, J. A. J., Hall, M. S., Rich, C. N., Gilliland, S. E., Ford, S. R., and Leach, F. R.,** Improved sensitivity of the bioluminescent determination of numbers of bacteria in milk samples, *J. Food Prot.,* 51, 949, 1988.
24. **Van Crombrugge, J., Waes, G., and Reybroeck, W.,** The ATP-F test for estimation of bacteriological quality of raw milk, *Neth. Milk Dairy J.,* (in press).
25. **Griffiths, M. W. and Phillips, J. P.,** The application of the pre-incubation test in commercial dairies, *Aust. J. Dairy Technol.,* 41, 71, 1986.
26. **Griffiths, M. W., Phillips, J. D., and Muir, D. D.,** Methods for rapid detection of post-pasteurization contamination in cream, *J. Soc. Dairy Technol.,* 37, 22, 1984.

27. **Phillips, J. D. and Griffiths, M. W.,** Bioluminescence and impedimetric methods for assessing shelf-life of pasteurized milk and cream, *Food Microbiol.,* 2, 39, 1985.
28. **Waes, G. and Bossuyt, R.,** A rapid method to detect postcontamination in pasteurized milk, *Milchwissenschaft,* 36, 548, 1981.
29. **Waes, G. and Bossuyt, R.,** Usefulness of the benzalkon-crystal-violet ATP method for predicting the keeping quality of pasteurized milk, *J. Food Prot.,* 45, 928, 1982.
30. **Sharpe, A. N., Woodrow, M. N., and Jackson, A. K.,** Adenosine triphosphate (ATP) levels in foods contaminated by bacteria, *J. Appl. Bacteriol.,* 33, 758, 1980.
31. **Stannard, C. J. and Wood, J. M.,** The rapid estimation of microbial contamination of raw meat by measurement of adenosine triphosphate, *J. Appl. Bacteriol.,* 55, 429, 1983.
32. **Kennedy, J. E. and Oblinger, J. L.,** Application of bioluminescence to rapid determination of microbial levels in ground beef, *J. Food Prot.,* 48, 334, 1985.
33. **Baumgart, J., Fricke, K., and Huy, C.,** Schnellnachweis des Oberflächenkeimgehaltes von Frischfleisch durch Biolumineszenz-Verfahren, *Fleischwirtschaft,* 60, 266, 1980.
34. **Littel, K. J. and LaRocco, K. A.,** Bioluminescent standard curves for quantitative determination of yeast contaminants in carbonated beverages, *J. Food Prot.,* 48, 1022, 1985.
35. **Littel, K. J., Pikelis, S., and Spurgash, A.,** Bioluminescent ATP assay for rapid estimation of microbial numbers in fresh meat, *J. Food Prot.,* 49, 18, 1986.

Chapter 4

THE PREPARATION AND PROPERTIES OF IMMOBILIZED FIREFLY LUCIFERASE FOR USE IN THE DETECTION OF MICROORGANISMS

M. F. Chaplin

TABLE OF CONTENTS

I.   Introduction 82

II.   Immobilization of Firefly Luciferase 82

III.   Preparation of a Bacterial Cellulose Support 83

IV.   Coupling of Luciferase to Cellulose 83

V.   Results of Immobilization Methods 86

VI.   Reaction Kinetics of Immobilized Luciferase 88

VII.   Effect of Diffusion on the Reaction Kinetics 92

VIII.   Concluding Discussion 94

Acknowledgment 94

References 94

I. INTRODUCTION

Over the years there have been many methods suggested for the analysis of ATP. Such methods have received close attention from microbiologists as the concentration of viable microorganisms is closely related to that of the ATP they contain and this endogenous ATP may be released by simply lysing the cells using various reagents.[1] Due to the complexity of the intracellular media containing such ATP, enzymatic methods of analysis are clearly indicated in order to ensure specificity in the response. The sensitivity of colorimetric assays enables the detection of ATP down to about 1 nmol. This limit may be reduced somewhat using coupled assays with substrate recycling but at the cost of lowered selectivity and the increased prospect of interference.[2] A much more sensitive process involves the luminescence reaction catalyzed by firefly luciferase. This can be used for the analysis of ATP down to femtomolar concentrations and with a linear response up to micromolar concentrations. It is not surprising, therefore, that this enzyme has become established as a reagent of choice for the specific analysis of low concentrations of ATP.[1,3-8]

The firefly luciferase assay has undergone a great deal of research and development in recent years.[9] Most of this has concerned the development of detailed and reproducible protocols for the assay and recipes maximizing the stabilization of the reagents.[10] The principal target of this work has been to reduce the rapid decline in light output (i.e., rate of reaction) that otherwise occurs during reaction and to increase the reproducibility of the assay. A relatively small number of workers have investigated the potential of immobilization of the firefly luciferase for use in these assays. This chapter is presented in order to analyze the results of these immobilizations, to introduce some previously unpublished data on the methods available for immobilization of luciferase, and to discuss the potential of these methods.

II. IMMOBILIZATION OF FIREFLY LUCIFERASE

There are three main reasons why immobilization may be of interest; economic, kinetic, and operational. Firefly luciferase is an expensive enzyme. Although it is relatively easy to purify and occurs at a high concentration within the light-emitting organs of the firefly, the only present source of the enzyme is the wild firefly which has to be caught by hand. The enzyme is, therefore, quite expensive at about £5000 U^{-1}. When the soluble enzyme is used in an assay, enough enzyme must be present in order that the reaction may progress reasonably rapidly. However, this means that substantial enzyme activity is left at the end of the reaction which cannot be economically retrieved and, consequently, is discarded. Immobilized enzymes may generally be reused with substantial savings in cost. Every time they are reused their cost is effectively reduced; 100 reuses reduces the unit cost by 99% which may easily compensate for the extra expense and activity losses involved in the immobilization process. It has also been established that immobilization sometimes causes substantial stabilization of the catalytic activity of the enzyme. Any such stabilization serves to increase the effectiveness of this economic argument. Enzymes are generally immobilized at high concentrations which enable high rates of reaction. This normally reduces the time necessary for assays with consequent savings in the cost of the analyst's time. Assays using luciferase are rapidly performed, however, as the light output gives the rate of reaction directly and there is no need for the time-consuming determination of the rates of change of substrate or product concentrations.

Immobilization is an important process as it may be used to change the kinetic constants of an enzyme, particularly its K_m, k_{cat} (V_{max}), pH optimum, susceptibility to inhibitors and denaturants, and the rate of inactivation by the physical conditions. As luciferase is considerably affected by inhibitors and rapidly loses activity away from specially formulated environments, it presents a prime candidate for kinetic manipulation by immobilization. Additionally, immobilization may reduce the decline in light output during reaction.

Immobilization of an enzyme can have a profound effect on the way in which the enzyme may be used. This is particularly relevant to analytical enzymes where the possibility of using them in a biosensor device is indicated. At present, luciferase is used in conjunction with heavy and expensive luminescence detectors but it is likely that advances in electronics will enable small and cheap portable sensors to be generally available in the near future. In order to maximize their effectiveness, reusable immobilized luciferase probes must be developed in order to replace present liquid reagent preparations for use under field conditions.

III. PREPARATION OF A BACTERIAL CELLULOSE SUPPORT

Bacterial cellulose was prepared from *Acetobacter xylinus* (NCTC 8621) grown under aerobic conditions in a media consisting of 5% w/v sucrose, 0.5% w/v yeast extract, 2% v/v ethanol, 0.25% v/v glacial acetic acid, and between 2 and 14% w/v polyethylene glycol (PEG 6000). The cellulose mat produced after 5 to 21 days was either homogenized or used vdirectly as a membrane. In both cases it was washed with dilute NaOH (2 *M*) and water before drying by acetone treatment or lyophilization. The presence of PEG considerably increased the rate of formation of the cellulose pad. Higher concentrations at greater than 2% w/v did not appear to have any additional influence, however. It was noted that it was important to subculture from media containing at least 2% PEG if a cellulose mat was to be produced at the higher PEG concentrations, otherwise a low yield of a dense lumpy product formed. The PEG was initially included in order to increase the porosity of the growing cellulose pad but its use did not seem to increase, to any significant extent, the ability of the cellulose to bind luciferase. The time of harvest did appear important, with 5 days growth, which was just sufficient to give an acceptable membrane, being optimal. The bacterial cellulose was either used directly or homogenized before use. The membranes formed from this cellulose were remarkably strong and did not disintegrate at any stage in use or under any of the reaction conditions examined. In particular, they were stable and resistant to drying, allowing the retention of activity when bound to luciferase, even after air drying and 5 days' storage dry in the light at room temperature (20°C).

IV. COUPLING OF LUCIFERASE TO CELLULOSE

The luciferase was coupled to the cellulose using cyanogen bromide[11] or a chloroformate.[12] In the latter case, 0.3 g of the dry cellulose was suspended in 10 ml of dry pyridine. The suspension was placed in an ice bath, and the solution of 1 g of nitrophenyl chloroformate or trichlorophenyl chloroformate in 4 ml of dry acetone was added drop by drop. The mixture was incubated for 40 min at 0°C in an ultrasonic bath. The modified cellulose was washed with acetone, methanol, a mixture of water with methanol (50:50 v/v) and, finally, with a large amount of water. A suspension of 15 mg of dessicated firefly lanterns in 5 ml of 0.4 *M* sodium acetate buffer containing 0.05 *M* tris acetate, pH 8.0, and 3 m*M* MgATP was added to the washed preparation of cellulose. The mixture was incubated for 20 h at 0°C. A solution of glycine (final concentration 0.2 *M*) and dithiothreitol (10 m*M*) was then added. The modified cellulose was washed ten times with a buffer containing 0.05 *M* tris acetate pH 8.0 and 10 m*M* dithiothreitol followed by twice with 0.05 *M* tricine buffer (pH 7.8) containing 10 m*M* dithiothreitol. The preparation of immobilized luciferase obtained was usually kept at 0°C in 0.05 *M* tricine buffer pH 7.8 containing 10 m*M* dithiothreitol and 25% (v/v) glycerol. A number of variations in this method were tried using a variety of coupling buffers replacing acetate, blocking agents replacing glycine and assay buffers (Table 1). Firefly luciferase was also coupled noncovalently simply by its addition to the washed bacterial cellulose followed by incubation for 2 to 18 h and thorough washing to remove the unbound enzyme.

TABLE 1

**The Effect of Coupling Buffer, Blocking Agent, and Assay Buffer on Firefly
Luciferase Immobilized on Bacterial Cellulose**

Conditions	Relative response
Standard conditions	100
0.1 *M* Tris acetate/0.5 *M* sodium acetate pH 9.0 coupling buffer	31
0.1 *M* Borate pH 8.0 coupling buffer	29
0.1 *M* Triethanolamine pH 8.0 coupling buffer	27
0.1 *M* Tricine pH 8.0 coupling buffer	22
0.1 *M* Bis-tris pH 8.0 coupling buffer	11
0.1 *M* Pyrophosphate pH 8.0 coupling buffer	0.04
0.1 *M* bicarbonate pH 8.0, 0.5 *M* NaCl coupling buffer	0.2
0.1 *M* Tyrosine blocking agent	60
0.1 *M* Tyramine blocking agent	74
0.2 *M* 4-Aminobenzoic acid blocking agent	3
0.2 *M* 1,6-Diaminohexane blocking agent	87
0.2 *M* D-Glucosamine blocking agent	7
0.2 *M* 6-Aminohexanoic acid blocking agent	76
0.2 *M* Tris blocking agent	75
0.2 *M* Glycine + ATP/luciferin/oxygen blocking agent	172
0.2 *M* Glycine + ATP/luciferin blocking agent	190
0.025 *M* Glycylglycine pH 7.8 assay buffer	89
0.025 *M* Tris acetate pH 7.8 assay buffer	76

Note: The standard conditions are 0.1 *M* tris acetate/0.5 *M* sodium acetate pH 8.0 coupling buffer, 0.2 *M* glycine
blocking agent, and 0.025 *M* tricine pH 7.8 assay buffer.

A comparison between the chloroformate-activated and nonactivated "blank" membranes showed that comparable activity was obtained using the "blank" (noncovalent) treatment. This suggested that very little, if any, activation was actually taking place, and the enzyme was simply being nonspecifically adsorbed onto the membrane. However, the kinetics for the noncovalent and covalent membrane preparations were noticeably different, covalently bound luciferase showing a far more gradual and extended rise to peak light output (Figure 1). Also, the untreated membrane had a very different appearance to the treated material. It was fluffier and quite thick while after activation it became very fine and smooth. Therefore, it is possible that the surface of the untreated membrane represents a more suitable substrate for noncovalent binding than the chloroformate-activated membrane. It was realized that it is not necessarily important that noncovalent attachment is taking place, as long as the final preparation exhibits high activity and stability. However, it was thought that it would be desirable if more covalent linkages could be formed as, generally, long term stability, and stability to varying operating conditions, are expected to be greater if the enzyme is covalently immobilized rather than if noncovalently adsorbed. Increased covalent attachment could be achieved by replacing the pyridine basic catalyst by 4-dimethylaminopyridine[13] and adding this catalyst to the membrane together with the chloroformate in acetone, in contrast to the previous protocol where the chloroformate was added to the membrane in the catalyst.

Cellulosic dialysis film was also crosslinked to luciferase by this chloroformate method. However, neither the untreated dialysis film, nor film treated with acid (1 *M* HCl) or alkali (1 *M* NaOH) for 3 days at room temperature appeared suitable for immobilization. In all cases, the film only contained traces of activity after coupling the luciferase. Pretreatment of the dialysis film with hydrogen peroxide, with and without catalase, or 10% sulfuric acid was also of no avail. Our failure to crosslink luciferase with the dialysis film can be explained by the fact that both treated and untreated film contained at least 500-fold less free accessible

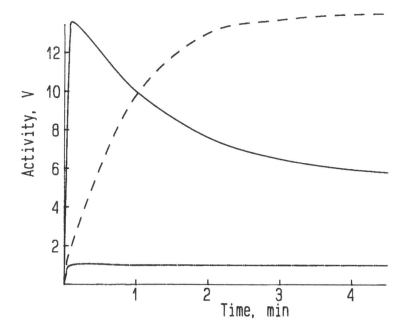

FIGURE 1. Luciferase-catalyzed reaction response. (—) Free soluble enzyme; (- - - - -) luciferase covalently immobilized on bacterial cellulose; (—·—·—·—) luciferase noncovalently immobilized on bacterial cellulose

TABLE 2
Activity of Some Immobilized Luciferase Preparations

Preparation	Accessible OH groups per mg dry weight	Relative response
Bacterial cellulose	5×10^{-7}	100
Cellulose (dialysis) film	10^{-9}	0
Dialysis film modified with D-glucosamine	10^{-8}	0
Dialysis film modified with 1-amino-5-pentanol	4×10^{-9}	0.7
Dialysis film modified with 1-amino-3-diethylamino-2-propanol	5×10^{-9}	0.4
Dialysis film modified with 1-amino-2,3-propanediol	2×10^{-9}	1.9

and, most importantly, reactive hydroxyl groups than bacterial cellulose. To increase the number of reactive hydroxyl groups, we treated the dialysis membranes consecutively with sodium periodate and amino alcohols. The Schiff bases formed were fixed by reduction with sodium borohydride. Films modified this way were able to react with 2×10^{-9} to 10^{-8} moles of chloroformate per milligram dry weight, allowing coupling to luciferase as above (Table 2). Attempts to immobilize luciferase by its direct reaction to the aldehyde groups generated in the cellulosic film by direct periodate treatment followed by borohydride reduction were unsuccessful. Immobilized luciferase obtained in this way exhibited very low activity. Attempts to use the periodate-oxidized and borohydride-reduced film for the chloroformate coupling were equally unsuccessful, although such treatment would be expected to open out the cellulose structure somewhat.

The observation was made that, during the periodate oxidative incubation of the cellulosic dialysis membranes, a thick deposition (thought to be consisting of sodium iodate crystals) formed, possibly retarding or preventing further reaction. In an attempt to prevent this possibility (1) lower concentrations of sodium periodate were used, (2) lower pHs were

TABLE 3
The Effect of the Treatment of Cellulose (Dialysis) Membrane During Periodate Oxidation on Final Immobilized Enzyme Activity and Crystal Deposition

Treatment (NaIO₄)	Crystal deposition	Response, mv mg⁻¹ (29 nmol ATP)
0.2 M, pH 7.0, 20°C, 1 h	+ + +	2.9
0.2 M, pH 7.0, 4°C, 1 h	+ + + + +	6.1
0.2 M, pH 7.0, 4°C, 4 h	+ + + + +	9.4
0.03 M, pH 7.0, 20°C, 20 h	+ +	0.9
0.2 M, pH 5.5, 4°C, 1 h[a]	+	2.3
0.2 M, pH 7.0, 4°C, 1 h[a]	+ +	2.7
0.2 M, pH 7.0, 20°C, 1 h[b]	−	3.1
0.2 M, pH 7.0, 20°C, 1 h followed by reduction, and then 0.005 M, pH 7.0, 20°C, 48 h	+ + + −	5.7
0.2 M, pH 5.5, 20°C, 1 h followed by reduction, and then 0.005 M, pH 5.5, 20°C, 48 h	+ −	1.7

[a] Solution changed every 10 min.
[b] Solution sonicated throughout.

examined, and (3) sonication during the reaction was tried. The results (Table 3) confirmed that incubation at 4°C with 0.2 M sodium periodate at pH 7.0 and without sonication produced a membrane which gave the highest activity after immobilization even though the deposition of crystals was the greatest under these conditions. It was decided to attempt to improve the reactivity of the dialysis membrane by double oxidation. The membrane was periodate oxidized and borohydride reduced as before to introduce some reactive diol groups. It was then oxidized again, under milder conditions, in order to convert these diols to aldehydes, followed by conversion to Schiff bases, and fixing by borohydride reduction. The immobilized luciferase produced from this doubly activated membrane showed a small increase in activity but this was thought to be too low relative to the increased effort needed to obtain it. In addition, the membrane became much more fragile.

V. RESULTS OF IMMOBILIZATION METHODS

Many different methods have been used to immobilize firefly luciferase (Table 4). It may be seen that the resultant preparations have generally been of low activity. Luciferase from two different species of firefly have been used in these studies, *Photinus pyralis*, from the U.S., in Western laboratories and *Luciola mingrelica*, collected in the Caucasus, in the U.S.S.R. These have been shown to be closely related enzymes with almost identical properties so it is reasonable to assume that conclusions drawn from the use of either may be applicable to the other.[14]

Optimization of the immobilization of firefly luciferase is more of a multivariate problem than many immobilization processes involving other enzymes as the native enzyme is quite unstable and easily inhibited and inactivated. All the reaction conditions, at each stage in the coupling, must be optimized before success can be realized. Additionally, the reaction kinetics are complex, with three substrates and several stable intermediates so immobilization may have a considerable influence on the resultant rate of reaction. There is also the possibility that immobilization may alter the rate-controlling step. Clearly for any process involving light emission, as in this case, the immobilization matrix must be fairly transparent in order to allow most of the photons released to be detected. The high-activity preparations are suitable for the detection of the ATP derived from bacterial suspensions at concentrations down to about 10^5 organisms per milliliter but are of borderline utility, without a preconcentration step, for use at the more generally required limit of 10^4 organisms per milliliter.

TABLE 4
Immobilized Firefly Luciferase Preparations

Support	Enzyme immobilization method	Activity[a]	Ref.
Porous glass	*P.p.* Glutaraldehyde/alkylamine	$10^{-8}\ M$	31
Porous glass	*P.p.* Glutaraldehyde/alkylamine	$10^{-8}\ M$	45
Sepharose 4B	*P.p.* CNBr activation[11]	$10^{-11}\ M$	31
Sepharose 4B	*P.p.* CNBr activation/ethylene diamine	200 fmol	46
Sepharose CL6B	*P.p.* CNBr activation	3 pmol	25
Nylon membranes	*P.p* Preactivated Biodyne® immunoaffinity membrane	2.8×10^{-10} M	39
Nylon tubes	*P.p.* Glutaraldehyde/alkylamine/adipic acid dihydrazide	240 mv m^{-1}	43
Nylon tubes	*P.p.* Glutaraldehyde/alkylamine/diaminohexane	400 mv m^{-1}	43
Collagen film	*P.p.* Hydrazine activated[44]	$10^{-11}\ M$	42
Sepharose 4B	*L.m.* CNBr activation	500	14
Ultradex (dextran)	*L.m.* CNBr activation	800	14
Cellulose film	*L.m.* CNBr activation	3 to 10	14
Ultragel (dextran/acrylamide)	*L.m.* CNBr activation	30 to 70	14
Sepharose 4B	*L.m.* CNBr activation	5 to 9×10^3	18
Albumin gel	*L.m.* Glutaraldehyde	0	18
Polyacrylamide gel	*L.m.* Entrapment in 16% gel	0	18
Polyacrylamide gel	*L.m.* Adsorbtion to 16% gel	0	18
AH-Sepharose 4B	*L.m.* Carbodiimide coupling	0	18
CH-Sepharose 4B	*L.m.* Carbodiimide coupling	0	18
CH-Sepharose 4B	*L.m.* Woodward reagent activation	0	18
Cellulose film	*L.m.* Cyanuric chloride activation	1 to 40	20
Cellulose film	*L.m.* CNBr activation	0.75	20
Cellulose film	*L.m.* 18% NaOH/cyanuric chloride	750	20
Cellulose film	*L.m.* Long term storage/cyanuric chloride	2500	20
Cellulose film	*L.m.* Periodate/cyanuric chloride	5000	20
Sepharose 4B	*L.m.* CNBr activation	5 to 10×10^3	20
Cellulose film	*P.p.* 4-Nitrophenyl chloroformate activation	0	47
Cellulose film	*P.p.* Periodate/reductive coupling[48]	0.01	47
Cellulose film	*P.p.* Periodate/reductive 1-amino-5-pentanol/4-nitrophenyl chloroformate	4	47
Cellulose film	*P.p.* Periodate/reductive 1-amino-3-diethyl-amino-2-propanol/4-nitrophenyl chloroformate	3	47
Cellulose film	*P.p.* Periodate/reductive 1-amino-2,3-propane-diol/4-nitrophenyl chloroformate	11	47
Cellulose film	*P.p.* Periodate/reductive D-glucosamine/4-nitro-phenyl chloroformate	0	47
Bacterial cellulose film	*P.p.* 4-Nitrophenyl chloroformate activation	642	47
Bacterial cellulose film	*P.p.* Trichlorophenyl chloroformate activation	89	47
Bacterial cellulose film	*P.p.* 4-Nitrophenyl chloroformate + substrates	1940	49
Bacterial cellulose film	*P.p.* 4-Nitrophenyl chloroformate/4-dimethy-laminopyridine	15000	49
Bacterial cellulose + plant cellulose	*P.p.* 4-Nitrophenyl chloroformate activation	100	49
Bacterial cellulose homogenate	*P.p.* 4-Nitrophenyl chloroformate activation	2340	50
Bacterial cellulose homogenate	*P.p.* 4-Nitrophenyl chloroformate + substrates	4906	47
Bacterial cellulose homogenate	*P.p.* 4-Nitrophenyl chloroformate + substrates but no oxygen	5398	47
Bacterial cellulose homogenate	*P.p.* CNBr activation	6674	50

[a] Where the specific activity is given in the literature, the units have been converted to mv mg^{-1}.[20] In other cases, the sensitivity is given together with the complete units.

TABLE 5
Storage Stability of Luciferase Immobilized onto Bacterial Cellulose

Storage conditions (0.025 *M* tricine pH 7.8)	Half-life (days)
No additions, 4°C	1
10 m*M* dithiothreitol, 4°C	12
0.5 m*M* dithiothreitol, 10% v/v glycerol, 1 m*M* EDTA, 4°C	40
0.5 m*M* dithiothreitol, 10% v/v glycerol, 1 m*M* EDTA, 1 μ*M* sodium pyrophosphate, 4°C	50
10 m*M* dithiothreitol, 25% v/v glycerol, 4°C	140
10 m*M* dithiothreitol, 0.5 m*M* EDTA, 20°C	3
10 m*M* dithiothreitol, 25% v/v glycerol, 0.5 m*M* EDTA, 20°C	21
Luciferase immobilized on cellulosic dialysis film stored in 10 m*M* dithiothreitol, 4°C	3

Although noncovalent methods show some success, the resultant preparations have fairly variable specific activities when small samples prepared with crude enzyme are assayed, indicating uneven binding efficiency. The most successful covalent methods involve the more hydrophilic porous supports linked through externally situated amino groups on the luciferase. Blocking the active site with substrates or products, during immobilization, increases the specific activity of these immobilized luciferase preparations, not unexpectedly as they will prevent the detrimental reaction of the necessary active site lysine,[15] and thiol,[16,17] residues.[18,19]

When cellulosic dialysis film was used as support for luciferase immobilization, we found that less activity could be achieved. This was in contrast to that reported elsewhere,[20] which we found impossible to reproduce. This may have been due to us not being able to reproduce their undefined and uncontrolled long term storage conditions. Dialysis-film-bound luciferase was generally less stable than that bound to bacterial cellulose (Table 5). A possible explanation of this is that the dialysis film treated with periodate is partially hydrolyzed resulting in the dissolution and release of bound luciferase. This appears likely as, after periodate treatment, the film becomes much more fragile.

VI. REACTION KINETICS OF IMMOBILIZED LUCIFERASE

A schematic diagram showing the current knowledge concerning the kinetics of firefly luciferase is shown in Figure 2.[14,18,21-25] There is some debate over whether the luciferase from *Photinus pyralis* is monomeric or dimeric; genetic cloning experiments indicate the former,[26,27] whereas traditional kinetics indicate the latter.[15,18] Luciferase from *Luciola mingrelica* is thought to be dimeric. The kinetic evidence is substantial, however, and at the present time may not be safely ignored.

A typical luciferase-catalyzed reaction response is shown in Figure 1. The peak light output is reached in approximately 4 s and then the light output declines. A similar picture has been observed by many authors using pure or crude luciferase preparations. Increasing ATP concentrations give rise to more rapid and deeper declines in the course of the reaction.[24,28] These kinetics may be explained, under circumstances where a large loss of substrate due to its utilization is not responsible, by the formation of a slow dissociating, and inactive, complex of luciferase with the reaction product, oxyluciferin (i.e., E.Lox in Figure 2). The increase in the rate of decline in the light output can then be explained by the higher rate of reaction product formation at higher ATP concentrations. Several approaches have been put forward for stabilizing the luciferase response. Arsenate and pyrophosphate have stabilization effects.[24,29,30] It has been suggested that these compounds prevent light output decline by reducing the reaction rate but it is more probable that they increase the rate of dissociation of the reaction product, E.Lox Sulfate has been shown to have an effect similar to that of arsenate.[28]

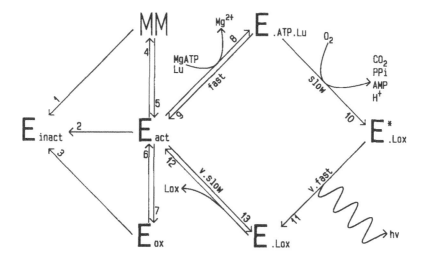

FIGURE 2. Firefly luciferase kinetics. The symbols have the following meanings: Eact, active luciferase dimer; Eox, oxidized luciferase; M, luciferase monomer; Einact, inactive (denatured) luciferase; Lu, D-luciferin; Lox, oxyluciferin; E*.Lox energetically activated luciferase-oxyluciferin complex; PPi, inorganic pyrophosphate; hv, light output. Inactivation of the luciferase takes place through the pathways 4 + 1, 2, or 7 + 3. Reaction takes place through the pathway 8 + 10 + 11 + 12. The equilibria 8 + 9 and 12 + 13 are rapidly or very slowly attained, respectively, as indicated. The activating (+) and inhibitory (−) influence of various reagents or conditions on the reactions have been determined to be: immobilization; 4 (−), 8(−), 10(−), 12(−); ammonium sulfate, 9(+), 12(+); glycerol; 5(+), 9(+), 12(+); dithiothreitol; 6(+), 10(+); molecular oxygen; 1(+), 7(+); pyrophosphate; 10(+); Triton® X-100 detergent; 12(+); bovine serum albumin; 2(−); hydrophobic surfaces 2(+).

The effect of glycerol depends on whether the enzyme is free or immobilized. Glycerol prevents the inactivation of diluted solutions of luciferase, apparently activating it tenfold (Figure 3). In the presence of glycerol the rate of luciferase reaction is linearly related to the luciferase concentration, whereas in its absence substantial inactivation occurs in dilute solutions (Figure 4). Glycerol also prevents the decline in light output in the course of the reaction due to its effect on the enzyme-inhibitor complex, E.Lox. Glycerol stabilizes immobilized luciferase but should not be present during ATP assay as it greatly inhibits its activity.

Ammonium sulfate also reduces the rate of light output decline although at a cost of some reduction in the rate of reaction; this inhibition probably being due to the increase in ionic strength.[31] Dithiothreitol only activates the luciferase slightly in the absence of ammonium sulfate but greatly protects luciferase in its presence (Figure 5). Despite the failure of ammonium sulfate to inhibit luciferase substantially in the presence of dithiothreitol, it is still able to prevent the decline in the rate of reaction. These results indicate that the stabilization caused by ammonium sulfate to the luciferase response is due not only to its inhibitory effect but to its effect on the dissociation of the luciferase reaction product, E.Lox. In contrast to the action of glycerol, these effects on soluble luciferase are equally applicable to the immobilized enzyme. It is recommended, therefore, that immobilized luciferase be stored in the presence of 25% v/v glycerol and 10 mM dithiothreitol but reacted in the presence of 50 mM ammonium sulfate and 10 mM dithiothreitol. It is known that dithiothreitol rapidly oxidizes in solution within a few hours if not protected from molecular oxygen.[32] However, as it protects luciferase over a much longer period than this it is clear that it must form disulfides with it that are equally protective. The active enzyme may be regenerated when fresh dithiothreitol is added just prior to ATP analysis.

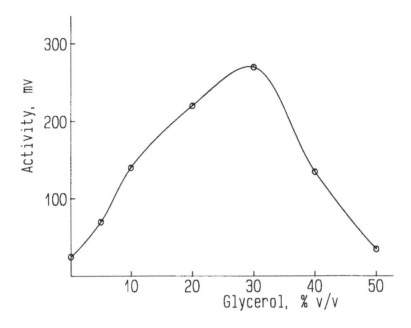

FIGURE 3.   Activation of low concentrations of firefly luciferase (2 μg ml⁻¹) by glycerol.

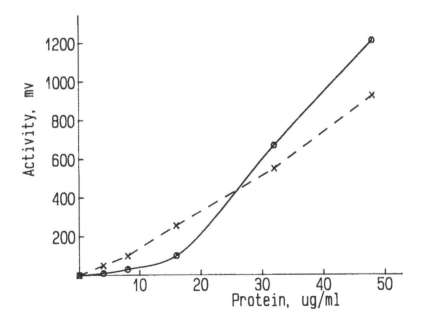

FIGURE 4.   The dependence of luciferase activity on its concentration. (-o-o-) Without added glycerol; (x- - -x) in the presence of 25% (v/v) glycerol.

Glycerol and ammonium sulfate are expected to act in antagonistic ways on both protein-protein and protein-ligand interactions. Increased salt concentrations increase the strength of hydrophobic interactions while reducing the strength of salt links. Glycerol is expected to reduce the strength of hydrophobic interactions while increasing the strength of salt links. As protein-protein and protein-ligand interactions are generally a combination of these two effects, glycerol and ammonium sulfate may have relatively different effects on the various

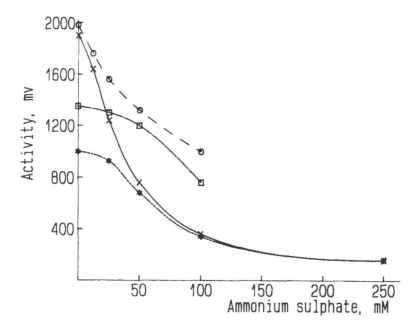

FIGURE 5. The dependence of luciferase activity on ammonium sulfate concentrations; 0.025 M tricine, pH 7.8, 10 mM MgSO$_4$, 15 μM D-luciferin, 40 μg ml^{-1} crude luciferase, 1 μM ATP. x—x peak reaction rate, at 4 s; *---* reaction rate after 4 min; o---o peak reaction rate in the presence of 5 mM dithiothreitol, at 4 s; □---□ reaction rate in the presence of 5 mM dithiothreitol, after 4 min.

associative and dissociative processes which occur during the luciferase reaction. In particular, it appears that as inactivation is noticed at low luciferase concentrations and, as this is most likely to be due to dissociation of the luciferase dimer prior to denaturation, glycerol must prevent this dissociation. It is interesting to observe that the activity decline occurs at and below about 3 × 10^{-8} M luciferase. This is in agreement with the dissociation constant (K$_d$ = 1.3 × 10^{-8} M) proposed elsewhere[18] for the *Luciola mingrelica* enzyme. It also follows from the preceeding discussion that the protein-ligand interactions during the luciferase reaction process are finely balanced between hydrophobicity and salt links and that interference with the strength of either causes an increase in the rate of dissociation.

Although bovine serum albumin has been found to be beneficial to the stability of soluble luciferase,[33] it appears to have no such effect on the immobilized enzyme. Indeed, its presence leads to reduced specific activities due to its successful competition for potential luciferase binding sites.

Immobilization of the luciferase stabilizes the enzyme as well as preventing the decline in light output during the course of the reaction. In our laboratory immobilized luciferase could be stored at 4°C for a month and at room temperature (20°C) for at least several days without loss of activity, a result that has been noted elsewhere.[34] Immobilized luciferase could be used many times for ATP analysis without loss of its activity. Clearly, immobilization may produce its stabilizing effect by preventing the dimeric dissociation which may otherwise lead to its rapid inactivation (see Figure 1). Also the immobilization support physically prevents the immobilized luciferase from contacting the surface of the vial used in the analysis so preventing a known[24,35] mechanism for its inactivation. Immobilization to hydrophobic supports (e.g., polyacrylamide[18]) however, may reintroduce this problem to the detriment of the stability of the immobilized enzyme. It has been reported[2] that there is a large conformational change in the luciferase molecule on binding its substrates. It is to be expected that immobilization will result in somewhat stiffer, more inflexible, enzyme so reducing the binding of both substrates and products and producing lower reaction rates, in agreement with the observed kinetics. Because of the complex kinetics of luciferase such changes in the binding energies are not necessarily expected to result directly in changes in the Michaelis constants.

Under conditions where there are no diffusional restrictions on the reaction rate, the immobilization support can change the apparent pH optimum of bound luciferase. The use of charged supports results in changes in the value of K_m for the doubly negatively charged MgATP molecule. It may be reduced tenfold by use of positively charged DEAE- cellulose as the immobilization suppprt, at low enzyme loading.[36]

VII. EFFECT OF DIFFUSION ON THE REACTION KINETICS

One factor that must always be taken into account concerning the kinetics of immobilized enzymes is the possibility that the rate of reaction is limited by the rate of substrate diffusion to the enzyme prior to reaction and the rate of product diffusion away from the active site. This has received some comment with respect to immobilized luciferase over the last few years. Several papers[2,37] have noted that diffusional limitations appear to be partially responsible for the changed kinetics on immobilization whereas others[19,20] attempt to prove by theoretical reasoning that this factor is of little or no consequence. I hope to show here that these latter papers present a flawed argument.

There are three substrates that have to diffuse to the immobilized luciferase before reaction can occur: luciferin, MgATP, and molecular oxygen. Generally, the rate-controlling substrate will be the MgATP because of its low bulk concentration in typical assays; 10^5 bacteria ml^{-1} being equivalent to about 10^{-10} M MgATP. It is important that the diffusion of the reaction products is also taken into account as their movement may also be affected by mass transfer restrictions and localized high concentrations of products may give rise to product inhibition or reaction reversal. The effect, however, would be less important at low bulk ATP concentrations due to the restricted ability for the build-up of localized product concentrations. Indeed, the accumulation of pyrophosphate may be beneficial as it helps the regeneration of the luciferase after reaction by releasing the reaction products. The effect is unlikely to be of importance under the usual conditions of high enzyme loadings and low ATP concentrations when there would always be an excess of unreacted and hence uninhibited enzyme awaiting reaction.

The effect of the diffusion of substrates to the surface of an immobilized enzyme is best described by the dimensionless external substrate modulus (μ)[38] where

$$\mu = \frac{V_{max,A}\delta}{K_m Ds}$$

$V_{max,A}$ is the maximum rate of reaction per unit area, which is proportional to the bound enzyme concentration. The maximum value that this could have as an enzyme monolayer on a perfectly flat surface is about 10^{-12} mol luciferase cm^{-2} s^{-1}. K_m is the Michaelis constant under the experimental conditions. A variety of determinations of this value have been made, but it is unlikely to be much lower than 10^{-8} mol cm^{-3} and may be as high as 10^{-6} mol cm^{-3} for the immobilized luciferase. Ds is the diffusivity of MgATP. The MgATP must diffuse as an electrically neutral species by codiffusing with sodium, potassium, hydrogen, or extra magnesium ions. Because of this, its diffusivity is likely to be about 2×10^{-6} cm^2 s^{-1}. δ is the effective depth of the stagnant layer above the flat surface and depends on the hydrodynamic conditions. The flow over a flat surface in a stirred reactor causes δ to be relatively large but will be no greater than about 10^{-2} cm. Using these estimates the maximum value that the external substrate modulus may have is about 0.5. A value this low indicates that there will be very little effect on the rate of reaction due to substrate diffusion. This lack of effect can be mainly attributed to the low turnover number (about 0.02 to 0.1 s^{-1}) of the enzyme. More generally μ would be expected to be considerably smaller due to the lower immobilization efficiencies more commonly encountered.

The effect of diffusion into porous particles or membranes containing immobilized enzyme is quite different however. The dimensionless internal substrate modulus (ϕ) is defined differently dependent on the type of porous catalyst. For a membrane it is given by

$$\phi = t\left\{\frac{V_{max,v}}{K_mDs^{eff}}\right\}^{1/2}$$

$V_{max,v}$ is the maximum rate of reaction per unit volume. The maximum value that this could have is about 10^{-7} mol luciferase cm^{-3} s^{-1}. Ds^{eff} is the effective diffusivity of electroneutral MgATP within the pores which is likely to be somewhat lower than its diffusivity free in solution especially if either the porosity or tortuosity of the biocatalyst, or both, is significantly different from unity. t is the thickness of the membrane which is usually in the range 10^{-1} to 10^{-2} cm. Using the K_m value above, this ϕ equals about 100. A value this high indicates a considerable effect of internal mass transfer on the rate of reaction.

The effect of diffusion on spherical porous biocatalysts is rather less than that experienced by membranes, as the amount of enzyme contained is lower for a given surface area and the substrates encounter reduced amounts of enzyme on penetration into the biocatalyst.

A most important but often neglected effect is the combined action of internal and external diffusion restrictions. In this case the membrane can be treated as a surface but, clearly, contains far more enzyme than can reside on a flat surface. The V_{max} for a membrane 2 × 10^{-2} cm thick using the previous values for the enzyme content per unit volume is 10^{-9} mol cm^{-2} s^{-1} giving rise to a value for the dimensionless external substrate modulus of 500. This indicates a strong likelihood of external diffusional limitation to the reaction rate even at lower enzyme loading and higher effective values for the Michaelis constants.

An additional effect was first described by Blum and Coulet[37] where it appeared that initial saturation of immobilized luciferase with D-luciferin increased the effect of ATP mass transport limitation. This may be explained by partition causing the localized ATP concentration within a few nanometers of the luciferin-loaded active site to be lower than that further away, perhaps due to like-charge repulsion. Such an effect would reduce the ATP concentration gradient down which the ATP is delivered and slow down the reaction. Such partition effects have been noticed before where the luciferase has been attached to charged supports.[36]

Immobilized luciferase preparations generally have a longer transient period before maximum light emission is achieved (Figure 1).[21,37,39] This could be due to the immobilization process affecting the kinetics of the enzyme directly but is more likely due to the limiting diffusion of the MgATP into the porous biocatalyst at the beginning of the assay. For reasons explained above, such an effect would not be expected to be as pronounced, and indeed is not found, where the enzyme is immobilized solely to a nonporous surface.

In the case of luciferase immobilized to bacterial cellulose there appears to be an initial very rapid rise in light output lasting about 5 s followed by a more gradual rise for about a minute and a very stable plateau (Figure 1). This agrees with the view, expressed above, that the externally diffusion-limited reaction rate only becomes noticeable as the ATP diffuses into the porous biocatalyst.

Perhaps the most important facet of the control of an immobilized enzyme reaction by external mass transport is the consequent reduction in the influence of the enzyme kinetics on the rate of reaction. In the extreme case with immobilized luciferase, the rate of reaction will only depend on the rate of arrival of MgATP, in turn proportional to the bulk ATP concentration. The reaction rate would therefore be independent of the temperature, pH, and inhibitors so long as the controlling process remained as substrate diffusion. Clearly, this is an important factor in the investigation of improved ATP assays as many analyte solutions (e.g., treated urine samples containing lysed bacterial cells) are of extremely variable composition and contain inhibitory substances.[6,40] Also, temperature control is very important when soluble luciferase is used as a 1°C change in temperature can give rise to a 5% change in activity.[6,24] Removal of the need for such close temperature control would allow the development of cheaper and portable biosensor devices.

VIII. CONCLUDING DISCUSSION

One may ask why immobilized luciferase preparations have not yet been accepted by the rapid microbiological methods community. Although luciferase is an expensive enzyme only relatively small (μg) amounts are necessary in the ATP assay as the determination of light output is so sensitive. Much expense has been used in generating uniform stable preparations of the free enzyme which cost about £1 per assay sample. This price is clearly low enough for workers in clinical fields where even the high cost of luminometers and trained personnel are not major factors. If a portable biosensor was developed that would allow clinical analyses within doctors' surgeries, this economic argument might be put more severely to the test. However, the sensitivity of presently available photodiodes, necessary for such portable low-cost biosensors, is far less than that of existing photomultiplier tubes;[41] the more sensitive phototransistors having consequentially higher background "noise" and the much more sensitive avalanche photodiodes needing a more stable and higher voltage source. The possibility exists, however, that even the small signal available from a photodiode may be sufficiently electronically amplified to be of use so long as the background "noise" level is kept to low levels and is stable, thus allowing it to be electronically filtered out or balanced by similar but light-shielded devices. Stable light output enables such devices to attain higher efficiencies by effectively integrating the light output. If and when such devices become available there will be a clear need for stable, immobilized firefly luciferase preparations.

ACKNOWLEDGMENT

I thank the Department of Health and Social Security for help with this project.

REFERENCES

1. **Stanley, P. E.**, Extraction of adenosine triphosphate from microbial and somatic cells, *Methods Enzymol.*, 133, 14, 1986.
2. **Leach, F. R.**, ATP determination with firefly luciferase, *J. Appl. Biochem.*, 3, 473, 1981.
3. **Kricka, L. J.**, Clinical and biochemical applications of luciferases and luciferins, *Anal. Biochem.*, 175, 14, 1988.
4. **Thore, A., Ansehn, S., Lundin, A., and Bergman, S.**, Detection of bacteriuria by luciferase assay of adenosine triphosphate, *J. Clin. Microbiol.*, 1, 1, 1975.
5. **Thore, A.**, Luminescence in clinical analysis, *Ann. Clin. Biochem.*, 16, 359, 1979.
6. **Lundin, A.**, Analytical applications of bioluminescence: the firefly system, in *Clinical and Biochemical Luminescence*, Kricka, L. J. and Carter, T. J. N., Eds., Marcel Dekker, New York, 1982, 43.
7. **Ansehn, S., Lundin, A., Nilsson, L., and Thore, A.**, Detection of bacteriuria by a simplified luciferase assay of ATP, in *Proc. Int. Symp. on Analytical Applications of Bioluminescence and Chemiluminescence*, Schram, E. and Stanley, P., Eds., State Printing and Publishing, Westlake Village, CA, 1979, 438.
8. **Hanna, B. A.**, Detection of bacteriurea by bioluminescence, *Methods Enzymol.*, 133, 22, 1986.

9. **Stanley, P. E., McCarthy, B. J., and Smither, R.,** *ATP Luminescence: Rapid Methods in Microbiology,* Soc. Appl. Bacteriol., Tech. Ser., Vol. 26, Blackwell, Oxford, 1989.

10. **Leach, F. R. and Webster, J. J.,** Commercially available firefly luciferase reagents, *Methods Enzymol.,* 133, 51, 1986.

11. **Axen, R., Porath, J., and Ernback, S.,** Chemical coupling of peptides and proteins to polysaccharides by means of cyanogen halides, *Nature,* 214, 1302, 1967.

12. **Drobnik, J., Labsky, J., Kudivosrova, H., Audek, V., and Svec, F.,** The activation of hydroxy groups of carriers with 4-nitrophenyl and N-hydroxysuccinimidyl chloroformates, *Biotech. Bioeng.,* 24, 487, 1982.

13. **Miron, T. and Wilchek, M.,** Activation of trisacryl gels with chloroformates and their use for affinity chromatography and protein immobilisation, *Appl. Biochem. Biotechnol.,* 11, 445, 1985.

14. **Ugarova, N. N., Brovko, L. Y., Filippova, N. Y., and Berezin, I. V.,** Immobilized firefly luciferase and its analytical application, in *Microbial Enzyme Reactions: Proc. 5th Joint US/USSR Conf. Microbial Enzyme Reaction Project US/USSR Joint Working Group Prod. Substances Microbiol. Means,* Weetall, H. H. and Bungay, H. R., Eds., American Society of Microbiology, 1980, 215.

15. **Lee, R. T., Denburg, J. L., and McElroy, W. D.,** Substrate-binding properties of firefly luciferase. II. ATP-binding site, *Arch. Biochem. Biophys.,* 141, 38, 1970.

16. **Rajgopal, S. and Vijayalakshmi, M. A.,** Firefly luciferase: purification and immobilization, *Enzyme Microb. Technol.,* 6, 482, 1984.

17. **Alter, S. C. and DeLuca, M.,** The sulfhydryls of firefly luciferase are not essential for activity, *Biochemistry,* 25, 1599, 1986.

18. **Ugarova, N. N., Brovko, L. Y., and Kost, N. V.,** Immobilization of luciferase from the firefly *Luciola mingrelica* — catalytic properties and stability of the immobilized enzyme, *Enzyme Microb. Technol.,* 4, 224, 1982.

19. **McElroy, W. D. and DeLuca, M. A.,** Firefly luminescence, in *Chemi- and Bioluminescence, Clinical and Biochemical Analysis,* Vol. 16, Burr, J. G., Ed., Marcel Dekker, 1985, 387.

20. **Ugarova, N. N., Brovko, L. Y., and Beliaieva, E. I.,** Immobilization of luciferase from the firefly *Luciola mingrelica:* catalytic properties and thermostability of the enzyme immobilized on cellulose films, *Enzyme Microb. Technol.,* 5, 61, 1983.

21. **Aflalo, C. and DeLuca, M.,** Continuous monitoring of adenosine 5'-triphosphate in the microenvironment of immobilised enzymes by firefly luciferase, *Biochemistry,* 26, 3913, 1987.

22. **McElroy, W. D. and DeLuca, M. A.,** Firefly and bacterial luminescence: basic science and applications, *J. Appl. Biochem.,* 5, 197, 1983.

23. **Brovko, L. Y., Ugarova, N. N., Vasiléva, T. E., Dombrovsk, A., and Berezin, I. V.,** Use of immobilized firefly luciferase for quantitative determination of ATP and enzymes that synthesize and destroy ATP, *Biochemistry (Eng. Trans. Biokhimiya) (U.S.A.),* 43, 633, 1978.

24. **Schram, E. and Janssens, J.,** Effect of dilution and other factors on the behaviour of standard firefly luciferase reagents, *Biolumin. Chemilumin., Proc. Int. Biolumin. Chemilumin. Symp., 4th,* John Wiley & Sons, Chichester, England, 1987, 539.

25. **Kricka, L. J., Wienhausen, G. K., Hinkley, J. E., and DeLuca, M.,** Automated bioluminescent assays for NADH, glucose 6-phosphate, primary bile acids and ATP, *Anal. Biochem.,* 129, 392, 1983.

26. **DeLuca, M.,** Firefly luciferase: mechanism of action, cloning and expression of the active enzyme, *J. Biolumin. Chemilumin.,* 3, 1, 1989.

27. **Gould, S. J. and Subramani, S.,** Firefly luciferase as a tool in molecular and cell biology, *Anal. Biochem.,* 175, 5, 1988.

28. **DeLuca, M., Wannlund, J., and McElroy, W. D.,** Factors affecting the kinetics of light emission from crude and purified firefly luciferase, *Anal. Biochem.,* 95, 194, 1979.

29. **Schram, E.,** Fundamental aspects of bioluminescent reactions used in clinical chemistry, in *Luminescent Assays: Perspectives in Endocrinology and Clinical Chemistry,* Serio, M. and Pazzagli, M., Eds., Raven, New York, 1982, 1.

30. **Lundin, A.,** Applications of firefly luciferase, in *Luminescent Assays: Perspectives in Endocrinology and Clinical Chemistry,* Serio, M. and Pazzagli, M., Eds., Raven, New York, 1982, 29.

31. **Worsfold, P. J. and Nabi, A.,** Bioluminescent assays with immobilized firefly luciferase based on flow injection analysis, *Anal. Chim. Acta,* 179, 307, 1986.

32. **Stephens, R., Stephens, L., and Price, N. C.,** The stabilities of various thiol compounds used in protein purification, *Biochem. Educ.,* 11, 70, 1983.

33. **Hall, M. S. and Leach, F. R.,** Stability of firefly luciferase in tricine buffer and in a commercial enzyme stabilizer, *J. Biolumin. Chemilumin.,* 2, 41, 1988.

34. **Ugarova, N. N., Brovko, L. Y., and Berezin, I. V.,** Immobilized firefly luciferase and its use in analysis, *Anal. Lett.,* 13, 881, 1980.

35. **Suelter, C. H. and DeLuca, M.,** How to prevent losses of protein by adsorption to glass and plastic, *Anal. Biochem.,* 135, 112, 1983.

36. **Marsden, D. J. G.**, The Immobilisation of Luciferase onto Solid Supports, Part II thesis, Exeter College, Oxford, 1986.
37. **Blum, L. J. and Coulet, P. R.**, Atypical kinetics of immobilized firefly luciferase, *Biotech. Bioeng.*, 28, 1154, 1986.
38. **Chaplin, M. F. and Bucke, C.**, *Enzyme Technology*, Cambridge University Press, 1990.
39. **Blum, L. J., Gautier, S. M., and Coulet, P. R.**, Luminescence fiber-optic biosensor, *Anal. Lett.*, 21, 717, 1988.
40. **Nichols, W. W., Curtis, G. D. W., and Johnston, H. H.**, The identity and properties of firefly luciferase inhibitors in urine, *Anal. Biochem.*, 114, 433, 1981.
41. **Marks, K., Killeen, P., Goundry, J., and Bunce, R.**, A portable silicon photodiode luminometer, *J. Biolumin. Chemilumin.*, 1, 173, 1987.
42. **Blum, L. J., Coulet, P. R., and Gautheron, D. C.**, Collagen strip with immobilized luciferase for ATP bioluminescent determination, *Biotechnol. Bioeng.*, 27, 232, 1985.
43. **Carrea, G., Bovara, R., Mazzola, G., Girotti, S., Roda, A., and Ghini, S.**, Bioluminescent continuous-flow assay of adenosine 5'-triphosphate using firefly luciferase immobilized on nylon tubes, *Anal. Chem.*, 58, 331, 1986.
44. **Coulet, P. R., Julliard, J. H., and Gautheron, D. C.**, A mile method of general use for the covalent coupling of enzymes to chemically activated collagen films, *Biotechnol. Bioeng.*, 16, 1055, 1974.
45. **Lee, Y., Jablonski, I., and DeLuca, M.**, Immobilization of firefly luciferase on glass rods: properties of the immobilised enzyme, *Anal. Biochem.*, 80, 496, 1977.
46. **Wienhausen, G. K., Kricka, L. J., Hinkley, J. E., and DeLuca, M.**, Properties of bacterial luciferase/NADH:FMN oxidoreductase and firefly luciferase immobilized onto Sepharose, *Appl. Biochem. Biotech.*, 7, 463, 1982.
47. **Chaplin, M. F., Harris, D., and Kozlov, I. A.**, Immobilisation of luciferase from fireflies by the chloroformate method. Report to the Department of Health and Social Security, 1987.
48. **Chaplin, M. F., Kozlov, I., and Barry, R. D.**, The preparation and properties of immobilised firefly luciferase membranes for use in the detection of bacteriuria, in *ATP Luminescence: Rapid Methods in Microbiology*, Soc. Appl. Bacteriol., Tech. Ser., Vol. 26, Stanley, P. E., McCarthy, B. J., and Smither, R., Eds., Blackwell, Oxford, 1989, 243.
49. **Chaplin, M. F. and Barry, R. D.**, Unpublished data, 1987.
50. **Chaplin, M. F. and Schofield, T.**, Unpublished data, 1985.

Chapter 5

BIOLUMINESCENCE APPLICATIONS IN BREWING

Robert Miller

TABLE OF CONTENTS

I.   Rapid Methods...98
    A.   Introduction 98
    B.   Microbial Spoilage in Brewing 98
    C.   Requirements..99
    D.   Microscopical Methods..100
        1.   Immunological 100
        2.   Membrane Filtration 100
    E.   Other Methodologies 101

II.   ATP Bioluminescence 101
    A.   Theory 101
    B.   Summary of Brewing Literature 102

III.   Bioluminescence at Wellpark 103
    A.   General Findings 103
    B.   Chlorhexidene as Extractant...104
    C.   Detection of Yeast and Lactobacilli in Beer...........................104
        1.   Detection Time Trials...104
        2.   Method 105
        3.   Results..106
    D.   Other Applications 107
    E.   Conclusion 108

References...108

I. RAPID METHODS

A. INTRODUCTION

In this chapter, I will discuss microbial spoilage as it can affect the brewing process, and the means of detecting it. I will concentrate on the potential of the rapid methods which have been developed in recent years, and in particular on one of the methods which has attracted most attention, ATP bioluminescence. Some of the published work will be referred to, and our own work at the Wellpark Brewery described in more detail.

The modern brewery has considerable advantages over its predecessors when it comes to maintaining high microbiological standards. Hygienic materials and design, cleaning-in-place (CIP) systems, and the increased use of pure yeast cultures have combined with the age-old high temperature processes to reduce the potential for introducing undesirable microorganisms, with their subsequent harmful effects. It might therefore be thought that pressure on the industry for further improvements in this direction would have lessened; in fact, this pressure has increased. Consumers now have higher expectations, and less tolerance of lapses in product quality. The demand of supermarkets, liquor stores, and the export trade, is for longer shelf-life. The introduction of more susceptible products containing little or no alcohol have reinforced this trend. Relatively static worldwide beer markets have increased competition for market share, and exacerbated fears of the effects of microbiological and other problems in trade.

B. MICROBIAL SPOILAGE IN BREWING

Spoilage potential during the brewing process varies greatly from stage to stage (see Figure 1). It is unlikely to occur during wort production, because high temperatures will normally intervene quickly enough to prevent it; on the other hand, cold temperatures during conditioning will slow the growth of any spoilage organisms present to an extent where problems will not occur. The periods where most attention is concentrated are throughout fermentation, and in final package. In the course of wort production, we are endeavoring, among other things, to supply a medium which will contain all the growth requirements for our chosen brewing yeast, so that a satisfactory fermentation results. Suitable fermentation temperatures are therefore chosen, depending upon the yeast's growth range, the type of product sought, and economic factors. In those circumstances, it is hardly surprising that other microorganisms find the conditions equally enjoyable. Even here, we have some advantages — some bacteria are inhibited by hop resins, and others by the decrease in pH and increase in alcohol during fermentation; some wild yeasts by the rapid onset of anaerobic conditions. Coliform bacteria grow in wort if given sufficient time without competition from healthy brewing yeast, but the most common bacterial contaminants are the remarkable, well-named *Obesumbacterium*, which co-exists and competes with pitching yeast, and *Acetobacter* spp. (see Table 1).

In the final package, there are fewer nutrients available, and alcohol is usually present, as well as hop resins, relatively low pH, and little air. However, there are a few organisms capable of causing spoilage in this fairly hostile environment. The last three groups of bacteria in Table 1 are relatively rare; *Zymomonas* causes spoilage of cask-conditioned ales, while *Pectinatus* and *Megasphaera* have been the subjects of a few reports in the international literature. However, I will consider only the two common spoilage groups, i.e., certain wild yeasts, many of them *Saccharomyces* species, and some species of the genera *Lactobacillus* and *Pediococcus*, closely related, and collectively referred to as lactobacilli.

The latter attracted less attention until recent years, due to their complex nutritional requirements, slow growth rates, tendency to change characteristics in culture, and the general difficulty of working with them. However, their importance as a spoilage hazard, especially to long shelf-life products, has rekindled interest. The main culprits are heterofermentative types, i.e., those which produce a variety of catabolites from carbohydrates in addition to lactic acid.

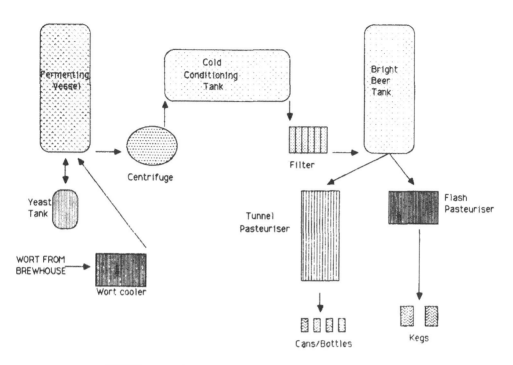

FIGURE 1. Typical brewing process. Wort cooling to final package.

TABLE 1
Brewing Spoilage Bacteria

Process stage	Potential spoilage bacteria
Early fermentation	*Obesumbacterium* Coliform types *Acetomonas* *Acetobacter*
Late fermentation, conditioning, and final package	*Lactobacillus* *Pediococcus* *Zymomonas* *Pectinatus* *Megasphaera*

Spoilage may be caused by a remarkably small number of contaminants. One detectable *Obesumbacterium* (i.e., one colony-forming unit) is capable of spoilage of at least 250 ml of unpitched wort. Spoilage of 250 ml of beer by one lactobacillus,[1] and of 75 ml by one *S. diastaticus*[2] have similarly been reported.

Earliest methods depended on light microscopy examinations, but the method is much too crude to detect small numbers of contaminants. Culture media have been the method of choice for many years. Yeasts and aerobic bacteria can be detected with little difficulty in 2 to 3 days whereas detection of some lactobacilli requires up to 7 days. Many growth media have been proposed,[3] and it is probable that no single medium will support the growth of all brewery lactobacilli.

C. REQUIREMENTS

There are various circumstances where we might wish to obtain more rapid microbiological information. These include trouble shooting, shelf-life prediction, and detection of procedural failure, e.g., contamination during yeast handling, or failure of pasteurization or sterile filtration. In a modern brewing plant, contaminant organisms are normally either absent, or present in only small numbers. Ideally, we want to immediately detect one such organism in the desired test volume, which may be up to several liters. It is also often essential to differentiate between live and dead cells. Methods must be reproducible and reliable, and should be inexpensive in terms of capital, running, and labor costs. The potential for automation is desirable, as is the survival of the microorganisms detected for further testing.

D. MICROSCOPICAL METHODS

This is a stringent list of demands, which did not begin to be addressed until the mid 1960s, when a number of studies based on microscopy began to appear, employing two approaches, immunological and micro-colony.

1. Immunological

One approach examined immunological methods for detecting wild yeasts in pitching yeast.[4] Two strains of wild *Saccharomyces* are injected into rabbits, and thereafter antisera prepared. Common antibodies are absorbed by treating with ale brewing strains. Fixed smears of pitching yeast are then treated with the antisera, and with a fluorescein which adheres only to the sites of antigen/antibody reaction. Wild yeasts fluoresce brightly when viewed under UV-light microscopy. Methods for *Lactobacillus*[5] and *Zymomonas*[6] contaminants were also developed. Immunological techniques are still in use, and yield results within the working day. However, they give no indication of viability, and the wild yeast method cannot be used for the more prevalent bottom-fermenting strains, which are apparently too similar to their wild relatives.

2. Membrane Filtration

The other approach dealt with filtered beer, and attempted to detect micro-colonies after a reduced incubation time. Samples are concentrated by membrane filtration, which virtually all subsequent methodologies have also utilized. The membrane filters used are usually cellulose-ester type, of 47 mm diameter, and 0.45 μ nominal pore size. After incubation, normally 18 to 24 h, a stain is applied. Several methods were proposed, the method of Richards,[7] based on the retention of the dye safranine by the micro-colony, being perhaps one of the most commonly used. In other versions, the membrane filters required to be dried, heated, and cleared, before examination by light microscopy.

Another type of micro-colony method utilized optical brighteners.[8] These methods enabled detection of a small number of yeasts; however, although some of them found a niche in brewery microbiological procedures, disadvantages remained—yeasts were much more readily detected than bacteria, and the organisms sometimes rendered nonviable.

More ambitiously, methods were developed for the detection of single viable cells on membrane filters. Molzahn and Portno[9] developed the work of Paton and Jones[10] for beer samples, based on the enzymic hydrolysis of fluorescein diacetate by living cells. Maximum esterase activity is induced by pre-incubation with sodium acetate, and after filtration the membrane is treated with the substrate, and a counterstain, e.g., methylene blue, to provide a darker background for the subsequent scan by incident-light microscopy. Viable cells fluoresce brightly. Yeasts are more readily detected than bacteria, due to their larger cell size and lesser risk of masking by beer solids. Other workers found a lack of correlation between fluorescence and cell viability, especially with heat-shocked cells,[11] and nonfluorescence of lactobacilli.[12] However, for the first time, events such as pasteurization failure could be detected in 2 h.

Rapid detection of lactobacilli has received much less attention. Kirsop and Dolezil[12] proposed the fluorochrome Euchrysine 2GNX with dark-field illumination, but did not consider incubation time for micro-colony formation, or infections of less than 50 organisms per liter.

It can be commented on all the foregoing methods that one of their principal practical disadvantages is the need for manual scanning of the smear or membrane filter. This is time consuming, especially when a number of samples is involved, and tends to cause strain on the operator.

More recently, with the development of image analyzers, interest has been shown in developing automated microscopical methods. Direct Epifluorescence Filter Technique (DEFT) combines acridine orange, which stains nuclear material, and an optical brightener (Tinopal AN) for the detection of bacteria. It is widely used in the dairy industry,[13] has been reported in the brewing literature,[14] and automated scanning is now available. Again, however, correlation with viability is not total, and the sample size which can be rapidly and automatically scanned is unacceptably small.

An automated micro-colony method based on two optical brighteners has been developed.[15] Tinopal CBS-X and Uvitex WG are incorporated into standard brewing media, and after incubation the Petri dish is inserted into an incident-light UV microscope connected to an image analyzer. Results are produced in 2 to 3 s, eliminating scanning problems. Recommended detection time for 100% recovery of aerobic organisms is 2 days by this method.

E. OTHER METHODOLOGIES

From the early mid-70s, approaches based on methodologies other than microscopy have been developed. One of the first was based on detection of C^{14}, either as precursors, e.g., lysine, taken up by metabolizing cells,[16] or as $^{14}CO_2$ from a labeled substrate such as glucose.[17] A liquid scintillation counter was used. It is sensitive enough to detect fairly small numbers of organisms, e.g., 2 *S. cerevisiae* in 24 h, but the cost of the instrumentation is high, and the use of isotopes in brewing laboratories has proved unpopular!

A method was developed[18] based on the ability of a microorganism to cause a decrease in the pH of a broth medium, of near-neutral pH and low buffering power. Detection times of 30 to 40 h were reported when 2/sample of a number of brewery organisms were inoculated. When unknown brewery samples were tested, the authors reported that results were sometimes obscured by the growth of incidental nonspoilage organisms,[8] a problem by no means confined to this approach.

Another effect caused by growth of microorganisms in a medium such as the above is a change in conductance or impedance due to the differing electrical characteristics of nutrients and metabolites. Highly automated instrumentation is available for handling in excess of 100 samples. Impedance may increase or decrease, depending on the nature of the contaminant. It has potential for detecting *Obesumbacterium* contamination in pitching yeasts at levels of 100 cells in 18 h,[19] and a method for detecting infection in forced beer samples has been developed.[20] However, the reliability and sensitivity of these approaches for routine use has been questioned, and the instrumentation is expensive.

Measurement of heat during growth in a defined medium by microcalorimetry has been proposed as a means of characterizing yeast strains.[21] Small numbers of organisms are, however, not detected.

II. ATP BIOLUMINESCENCE

A. THEORY

The foregoing summary brings me to the main topic of this chapter, the use of bioluminescence in the brewing industry, actual and potential. As the methodology is now well understood and widely known, it will suffice to say that in essence, a sample of live microbial cells is prepared, and an extractant added which has the dual role of releasing adenosine triphosphate (ATP) from the cells and inactivating intracellular ATPase enzymes. An enzyme extracted from fireflies, called luciferase, is added, which reacts with ATP, emitting light quantitatively, such that when other reactants are in excess, the light emitted is directly related to the amount of ATP present. Light output is measured on a photometer. As ATP content per cell is fairly constant, the number of cells may be calculated.

Interest in the methodology dates back to the 1940s. Progress was hampered for a considerable time by the impure nature of the firefly extracts then available.

B. SUMMARY OF BREWING LITERATURE

After workers in a number of fields had published studies in the early mid-1970s, Hysert et al.[22] published their work on detection and enumeration of brewery microorganisms. Using dimethyl sulfoxide as extractant, and the then best available firefly extract, measurement of the ATP contents of a range of typical brewery organisms was carried out on a liquid scintillation counter. These were in the range 0.5 to 5.0 fg per bacterial cell, and 100 to 500 fg per yeast cell, similar to other workers' findings.[23,24] The effects of age, and growth and temperature conditions were studied. Detection of as few as ten yeast cells was achieved. Their attempt to establish contamination levels in unpasteurized beers was unsuccessful, presumably because of interference from extracellular ATP. Beers have a signficant, variable extracellular ATP content, mainly due to their secretion by yeast during fermentation. If it is not removed, the sensitivity of the assay is diminished to some degree. Some workers have concluded that this effect is insignificant, whereas others have minimized it by treating samples with a washing procedure, an ATPase enzyme, or both. In a further study[25] the ratios of the three adenine nucleotides of various brewing yeasts were compared under different conditions. Stepwise enzymatic conversion of AMP and ADP to ATP was used. Significant interspecies differences were found, and proposed as a means of differentiation, especially between ale and lager yeasts. Levels of the nucleotides in various beers were measured, and found to be generally higher and more variable in ales. AMP levels were generally ten times higher than ADP and ATP. Changes in nucleotide concentrations during fermentations were measured, showing a rapid increase in the early stages, followed by a gradual fall. The ratio of nucleotides was used to derive the concept of Energy Charge. It is a tribute to the skill and perseverance of these workers that no significant increase in the sensitivity of the assay has since been achieved, despite improvements in reagents.

Miller et al.[26] proposed ATP measurement as a means to control the amount of yeast used for fermentation, known as the pitching rate. Diluted yeast samples were extracted with acetone, and assayed with a crude firefly extract. Accurate results were obtained over a wide range of yeast dilutions. This approach relies on the relatively constant ATP content per cell previously referred to.

Bioluminescence was one of the methods used by Ryder et al.[27] to investigate the relationship between yeast growth and glycolytic efficiency during fermentation. In addition to ATP, assays for NADH and NADPH were carried out using a photobacterial test system. By this means, the concentrations of various glycolytic intermediaries were measured.

The antibiotic nisin has been proposed as a bacteriocin to eliminate contamination by lactic acid bacteria. Waites and Ogden[28] developed an assay for nisin based on its ability to cause an increase in the concentration of extra- or intracellular ATP in *L. brevis.*

Kilgour and Day[19] developed a method for testing brewery rinse and swab samples. After membrane filtration of up to 1 l, samples were resuscitated, extracted, and assayed for ATP using the reagents of Lumac Ltd. One yeast, or 100 bacterial cells per milliliter were thus detected. Protocols for swab testing were recently developed by Simpson et al.[29] A new patented extractant, named BAX, was introduced. Correlation with plate counts was fairly good.

In his work on PET bottle infection, Avis[30] tested a variety of filter types and treatments, and recommended positively charged nylon filters of pore-size 1.2 μ. This enabled filtration of the entire 2-l contents. After overnight incubation in YM broth and ATPase treatment, good correlation was obtained for detection of small numbers (1 to 7) of yeast contaminants.

Simpson et al.[29] also described a method for testing unpasteurized beers. Positively charged nylon filters were used, extracellular ATP removal is by both washing and ATPase, and extraction is by BAX. Good correlation with plate counts was achieved, but problems were encountered in applying the method to pasteurized beers with few contaminants.

Dick et al.[14] tested beer samples containing three brewery microorganisms. Membrane-filtered samples were incubated in unstated broth media, no ATPase was used, and extraction was by boiling buffer. Under these conditions, calculated detection limits were 2 *S. cerevisiae*, 70 *L. brevis*, and 1200 *P. damnosus* in 48 h. These workers preferred the DEFT technique in terms of sensitivity and cost.

Concerning instrumentation, a comparative survey of the sensitivity of some of the commercially available equipment has recently been carried out.[31]

III. BIOLUMINESCENCE AT WELLPARK

A. GENERAL FINDINGS

Our own interest in this methodology began some eight years ago with a brief preliminary trial, and work has been more or less continuous since we purchased a photometer five years ago.

Despite its high-energy properties, the ATP molecule is extremely stable, and particularly heat resistant; it is practically unaffected by boiling, and surface sterilization with alcohol. Autoclaved or sterile disposable containers and utensils are not necessarily ATP free, and batches are checked before use by testing residual or rinse water with luciferase.

Main water contains ATP at concentrations high enough to interfere with microbial assays. All water used for reagent preparation is distilled and autoclaved.

We decided at an early stage that our beers required ATPase treatment; the enzyme used is of potato origin.

Most complex media, including wort and beer, cause considerable quenching of the luminescence signal. Table 2a and b shows the quenching effects of various media and ingredients. Where membrane filtration is employed, a washing stage minimizes this effect.

Luciferase reagents are another source of variability, from one manufacturer to another, and from batch to batch. Products showing greatest light output also show greatest background count. Some products are best held at room temperature for an hour or more between preparation and use, to reduce the background count. Deterioration due to age is somewhat better known. We find our solutions are stable for 1 month at −18°C, or 3 days at 4°C. There is a positive side to the stability of previously referred to ATP. Test solutions deteriorate only very slowly, and can be stored for several months at 4°C, and even longer at −18°C. Thus, the condition of the current luciferase can be tested against a known ATP solution, to provide daily assurance that satisfactory light detection is being achieved.

In general, we have found with our reagents that everything which could go wrong, has gone wrong at some time or another. The ubiquity of ATP, and the small quantities being assayed require regular vigilance. We therefore carry out daily checks for contamination on our luciferase, ATPase, and extractant, whether freshly prepared or not. Extractant *efficacy* is checked by measuring the light output from an approximately known number of microbial cells, and duplicate sterile beer samples are also tested. This approach to control samples was adopted as a satisfactory alternative to another often used system, that of adding an internal ATP standard to each sample, and measuring light output a second time. There are now several suppliers of high-quality bioluminescence reagents, and we find it satisfactory to prepare our luciferase, ATPase, and ATP according to manufacturer's directions.

TABLE 2
Quenching Effects of Brewing Media and Ingredients

2a. Whole broth media-ATP + luciferase + medium

Medium	Light output
Raka-ray	6.7
Wort	12.7
Nonalcoholic beer	16.6
Sucrose + yeast nitrogen base	38.3
MYGP	84
Lumacult	247
Distilled water	250

2b. Ingredients of Raka-Ray medium-ATP + luciferase + ingredient

Cause negligible quenching	Cause severe quenching	Insoluble when quencher's absent
Tween 80	Ammonium citrate	Dipotassium hydrogen phosphate
Manganese sulfate	Betain HCl	Magnesium sulfate
Maltose	Potassium aspartate	
Fructose		
Trypticase		
Yeast extract		
Liver concentrate		
Potassium glutamate		
N-acetyl glucosamine		

B. CHLORHEXIDINE AS EXTRACTANT

The only type of reagent which I wish to describe in some detail is the microbial extractant. The concepts affecting their choice has been discussed by Stanley.[32] We tested previously reported extractants, alcohol and trichloroacetic acid,[33] benzalkonium chloride,[34] acetone, and dimethylsulfoxide, but found them inconvenient at best, as they have to be removed before addition of luciferase, due to the effects of inappropriate pH, or quenching of the light signal. The most commonly used by far in the last few years is the Lumac product, NRB, and the two-stage BAX reagent is now available.

In our early trials, we used NRB, but have since changed to chlorhexidene, supplied as gluconate under the trade name Hibitane, (ICI). It is used at 0.3% in tris acetate buffer. Solutions are stable for 2 to 3 days and performance compares favorably with NRB. Cell breakdown occurs immediately, and the light output then remains constant for up to 1 h. Chlorhexidene also inhibits ATPase activity,[35] so that residual traces after washing do not cause breakdown of ATP in the sample. (However, attempts to simplify the methods below by adding the extractant directly to samples with ATPase still present were unsuccessful, possibly due to a difference in pH optima.) Chlorhexidene is a single-stage extractant, and much less expensive than NRB. It showed slightly better responses on membrane filters, although slightly poorer in unfiltered samples. Responses were slightly enhanced by rapidly heating the sample to 100°C, but not sufficiently to include this in the method.

C. DETECTION OF YEAST AND LACTOBACILLI IN BEER
1. Detection Time Trials

Initial tests in saline confirmed other workers' detection limits of ten yeast cells per sample in saline; in membrane-filtered beer samples, 20 is more realistic. 100 × these numbers of lactic acid bacteria were required.

As referred to above, very low levels of contamination by lactic acid bacteria could lead

TABLE 3
Detection of Yeast Spoilage of
Nonalcoholic Lager After 1 Day

Organism	Plate count on WLN		Light emitted (mV)
	(Initial)	(1 day)	
NCYC 324	8	1120	26.8
NCYC 324	8	1240	32.4
NCYC 324	3	590	30.9
NCYC 324	3	1490	69.9
NCYC 447	9	540	59.9
NCYC 447	9	560	57.2
Lager yeast	5	200	24.4
Lager yeast	5	310	34.6
Control	0	0	7.1
Control	0	0	7.4

to spoilage of final packaged products. Our aim was, therefore, to develop methods which reduced the incubation times for sterility testing as much as possible, while ensuring that 1 cfu per volume of product sampled could be confidently detected, even if the organisms are stressed. Few other studies on rapid methods have tested stressed organisms, but we concluded that the phenomenon could not be overlooked. It is likely to occur in samples where the microorganism is harbored in the product, which is pasteurized, or the product container, which is heat- or chemically treated. The effect on the lag phase of the organism may be considerable, and will not be known.

Detection of spoilage by forcing the product itself, and testing for growth after an adequate interval was found to be of limited value. It enables detection of yeast spoilage of alcohol-free, relatively nutrient-rich lager in 24 h (Table 3). However, for alcoholic beers and infections by lactobacilli, we were not particularly surprised to find that this approach offered only a small time saving over the standard forcing test, i.e., often longer than the current culture methods. A membrane filtration stage was therefore incorporated, followed by incubation in a broth growth medium. We then carried out a series of detection time trials. Cultures were prepared by inoculating yeasts and lactic acid bacteria into Wallerstein Laboratory Nutrient (WLN) or Raka-Ray broths, and incubating at 27°C for 24 or 72 h, respectively. They were then diluted in saline to the required concentrations, and added to pasteurized beers. If the organisms were to be heat stressed, approximately 10^4 cells were heated in beer to 53 to 54°C for 1 to 4 min, depending on observed thermal tolerance.

The medium now used for detecting lactic acid bacteria was developed in this laboratory, and referred to as Lactic Acid Bacteria medium (LAB; see Table 4). To render the medium as selective as possible for heterofermentative lactobacilli, a cocktail of antibiotics is added—10 ppm cycloheximide for inhibiting brewing, and some wild, yeasts, 0.15% 2-phenylethanol for Gram −ve bacteria, and 5 ppm vancomycin for most Gram +ve bacteria except beer spoilage lactobacilli.[36]

WL broth is used for yeasts. Similarly, it could be made more specific, by the inclusion of antibiotics to inhibit all bacteria.

2. Method

250 ml of beer is membrane filtered, and the membrane transferred facedown to a 55 mm Petri dish containing 2 ml of broth medium. WL broths are incubated aerobically for 24 h and LAB anaerobically for 72 h, both at 27°C. Then 1 ml of ATPase is added, and after 1 h at room temperature, the membrane is re-inverted into a filter holder, excess reagent

TABLE 4
Lactic Acid Bacteria Medium (LAB)

	g/l (unless otherwise stated)
Maltose	5
Fructose	5
Arabinose	5
Glucose	3
Arginine	2
Tryptose	20
Malt extract	5
Yeast extract	5
Liver concentrate	1
Calcium pantothenate	0.01
Dipotassium hydrogen phosphate	23.3
Sodium acetate	10
Magnesium sulfate	2
Manganese sulfate	0.5
Sodium pyruvate	0.8
Potassium aspartate	2
Potassium glutamate	2
Tween 80	10 ml

Note: Adjust pH to 5.7.

pipetted in, filtered, and washed with 100 ml water. The membrane is transferred to another sterile Petri dish, and 500 μl of chlorhexidine added. 200 μl is withdrawn into a cuvette, mixed with 100 μl luciferase, and the emitted light read in the photometer counting chamber. The instrument used is a Berthold® Biolumat LB 9500; emitted light is measured, as usual, by integration mode over 10 s after a 2-s delay. Manual injection is used, to maximize utilization of luciferase without contaminating it.

3. Results

In interpreting luminometry results as showing growth or no growth, a cut-off point had to be chosen. On the basis of our experience, samples showing more than three times the average of two control samples were taken as positive, i.e., growth had occurred.

All yeasts tested were detected overnight, approximately 18 h. Although most of the lactobacilli tested were detected in 2 days, some required 3 days at the lowest infection levels, so this was the incubation time chosen. All strains were detected at the lowest concentrations tested, whether in normal or heat-stressed condition.[37]

The anaerobic method is now in routine use, for 3-day release of long shelf-life export and low-alcohol products, with Raka-Ray plate counts as backup. The aerobic method has been used for trouble shooting (see Table 5a and b). As one would expect, most pasteurized beer samples do not contain any viable organisms and the remainder often only a very few, so that some random variation is to be expected. Therefore, duplicate samples of the standard aerobic method were also compared (Table 5c).

In the anaerobic series, more samples showed infection by luminometry, (21), than by plate count, (4). This indicates either that LAB is a superior medium, or, more likely perhaps, that the broth method is more sensitive, since many of these LAB broths were visibly turbid, and infection was detectable microscopically. Infection detected in the four plate counts was 1 to 4 cfu; two of these were also positive by luminometry.

Correlation (as determined by 95% confidence limits) between luminometry and standard methods was actually greater than between standard methods in duplicate; this emphasizes the difficulty of applying statistical treatment to this type of comparison, as discussed by Simpson et al.[29]

TABLE 5
Comparison of Bioluminescence and Plate Count Results

(a) Anaerobic method

	Bioluminescence	Plate count (P.C.)
Samples tested	414	414
Infection detected — total	21	4
Infection detected — biolum. only	19	—
Infection detected — P.C. only	—	2
Infection detected — both methods	2	2

(b) Aerobic method

	Bioluminescence	Plate count (P.C.)
Samples tested	66	66
Infection detected — total	2	1
Infection detected — biolum. only	1	—
Infection detected — P.C. only	—	0
Infection detected — both methods	1	1

(c) Duplicate plate counts

	Sample A	Sample B
Samples tested	85	85
Infection detected — total	22	24
Infection detected — A only	12	—
Infection detected — B only	—	14
Infection detected — both samples	10	10

Where a positive result is obtained, the first Petri dish used for incubating the sample is saline rinsed into a growth medium. It often contains a few residual organisms, thus providing cultures for further testing, and helping to counter the criticism sometimes made that the methodology is destructive. These methods confer considerable advantage over the present standard methods, especially the 7-day lactobacillus incubation. Stock may be cleared 4 days earlier, resulting in considerable cost savings.

The technique takes little longer to perform than existing methods, does not cause strain on the operator, and the cost of the reagents recommended above is acceptable. Sensitive photometers are relatively inexpensive.

D. OTHER APPLICATIONS

Some other proposed applications have already been referred to; it appears possible to assess total microbial load in beers or plant sterility samples at any process stage. We have carried out a little preliminary work on some of these.

ATP content is one of the methods suggested to give an indication of the suitability or otherwise of stored yeast for re-pitching, i.e., if the yeast's "vitality" was satisfactory. Serial dilutions were prepared to enable testing of a known number of yeast cells against an ATP standard, and the content per cell calculated. Our findings were that the small variations in ATP content due to the physiological condition of the yeasts were difficult to detect, being hampered by the day-to-day variability of reagents. It is perhaps worth adding that the other methods tested, i.e., oxygen uptake rate, glycogen content, and acidification power, equally failed to predict the condition of this brewery's yeasts as adequately as the timeless methylene blue stain. It should be pointed out that this is a different approach to that of Miller et al.,[26] where the aim was to quantify yeast via its ATP, rather than assess its condition.

In measuring high yeast concentrations, e.g., during fermentation, several sample dilutions are required, and also possibly ATPase and washing treatments. While feasible, this does not seem a rapid or economical way of measuring yeast concentration.

We have employed a variation of our packaged beer methods on unpasteurized bright beer. Five milliliters is membrane filtered, and since no incubation is required, ATPase is added immediately, and the sample processed as for pasteurized beer. Plate counts have shown that the ATP measured is preponderantly from yeast, with a negligibly small contribution from bacteria. Results are available in 1 h and results so far are promising.

Our own attempts at swab testing have come up against a problem mentioned by Simpson et al.,[29] variability in the ATP content of swabs, even from the same manufacturer and batch. However, Lumac now offers a kit including ATP-free swabs, and this type of test is now becoming part of routine quality control at a number of British breweries.

E. CONCLUSION

Three areas of progress appear likely in the near future. More sensitive photometers may be developed, and, perhaps more importantly, better and/or more inexpensive reagents. For instance, it is of considerable importance whether an ATPase removes 99.9 or 99.99% of extracellular ATP, in the food and dairy industries even more than in brewing. Detection of smaller numbers of organisms would then be enabled, perhaps one yeast cell. This may also be facilitated by testing larger samples.

Secondly, where immediate detection of small bacterial numbers is not possible, more specific media will be used. They will be based on nutrient-rich basal media, generally of near-neutral pH, possibly containing revivers for damaged cells. A short pre-incubation period to maximize resuscitation of damaged cells may be included. For the main incubation period, combinations of antibiotics will be incorporated to select for the desired organisms.

Thirdly, selective extractants may become available, enabling differentiation of, perhaps, yeast, Gram + ve, and Gram − ve bacteria.

The development of bioluminescence methodology has occurred at an increasing pace over the last 10 years in particular. Its introduction into routine usage had previously been hampered by a lack of good quality reagents and their high cost, problems which have now been largely overcome.

In this review, I have tried to show some of the difficulties and pitfalls in ATP bioluminescence, and that it does not supply the ideal for a rapid method of instant detection of a single microbial cell. I hope to have also demonstrated that it is at least as sensitive as the alternatives, fulfills many of our requirements, and that its potential range of applications in brewing microbiology is wide.

REFERENCES

1. **Brumsted, D. D. and Glenister, R. P.,** The viability of minimal populations of a *Lactobacillus* species in beer in relation to biological control limits for bulk pasteurisation, *J. Am. Soc. Brewing Chem.*, 12, 1963.
2. **Greenspan, R. P.,** The viability of minimal numbers of Saccharomyces diastaticus in beer, *J. Am. Soc. Brewing Chem.*, 109, 1966.
3. **Smith, C. E., Casey, G. P., and Ingledew, W. M.,** The use and understanding of media used in brewing microbiology, *Brewers Dig.*, October 12, 1987.
4. **Richards, M.,** Detection of yeast contaminants in pitching yeast, *Wallerstein Lab. Commun.*, 33, 11, 1970.
5. **Dolezil, L. and Kirsop, B. H.,** An immunological study of some lactobacilli which cause beer spoilage, *J. Inst. Brewing*, 81, 281, 1975.
6. **Dadds, M. J. S., Martin, P. A., and Carr, J. G.,** The doubtful status of the species Zymomonas anaerobia and Z. mobilis, *J. Appl. Bacteriol.*, 36, 531, 1973.

7. **Richards, M.,** Routine accelerated membrane filter method for examination of ultra-low levels of yeast contamination in beer, *Wallerstein Lab. Commun.,* 33, 97, 1970.

8. **Harrison, J. and Webb, T. J. B.,** Recent advances in the rapid detection of brewery micro-organisms and development of a micro-colony method, *J. Inst. Brewing,* 231, 85, 1979.

9. **Molzahn, S. W. and Portno, A. D.,** New advances in microbiological quality control, *Proc. Eur. Brewing Convention,* 15, 480, 1975.

10. **Paton, A. M. and Jones, S. M.,** The observation and enumeration of micro-organisms in fluids using membrane filtration and incident fluorescence microscopy, *J. Appl. Bacteriol.,* 38, 199, 1975.

11. **Chilver, M. J., Harrison, J., and Webb, T. J. B.,** Use of immunofluorescent and viability stains in quality control, *J. Am. Soc. Brewing Chem.,* 36, 13, 1978.

12. **Dolezil, L. and Kirsop, B. H.,** Detection of lactobacilli in brewing, *Proc. Long Ashton Symp.,* 159, 1973.

13. **Pettipher, G. L., Mansell, R., MacKinnon, C. H., and Cousins, C.,** Rapid membrane filtration-epifluorescent microscopy technique for the direct enumeration of bacteria in raw milk, *Appl. Environ. Microbiol.,* 423, 39, 1980.

14. **Dick, E., Wiedmann, R., Lempart, K., and Hammes, W. P.,** Rapid detection of microbial infection in beer, *Chem. Mikrobiol. Technol. Lebensm.,* 10, 37, 1986.

15. **Parker, M. J.,** The application of automated detection and enumeration of microcolonies using optical brighteners and image analysis to brewery microbiological control, *Proc. Eur. Brewing Convention,* 22, 545, 1989.

16. **Bourgeois, C., Mafart, P., and Thouvenot, D.,** Rapid method for detecting beer contaminants by means of radio-active labelling, *Proc. Eur. Brewing Convention,* 14, 219, 1973.

17. **Box, T. C. and Ferguson, B.,** Rapid estimation of small numbers of yeast by a radiotracer method, *J. Am. Soc. Brewing Chem.,* 33, 133, 1975.

18. **Harrison, J., Webb, T. J. B., and Martin, P. A.,** The rapid detection of brewery spoilage micro-organisms, *J. Inst. Brewing,* 390, 80, 1974.

19. **Kilgour, W. J. and Day, A.,** The application of new techniques for the rapid determination of microbial contamination in brewing, *Proc. Eur. Brewing Convention,* 19, 177, 1983.

20. **Evans, H. A. V.,** A note on two uses for impedimetry in brewing microbiology, *J. Appl. Bacteriol.,* 53, 423, 1982.

21. **Perry, A. E., Beezer, A. E., and Miles, R. J.,** Characterisation of commercial yeast strains by flow microcalorimetry, *J. Appl. Bacteriol.,* 54, 183, 1983.

22. **Hysert, D. W., Kovecses, F., and Morrison, N. M.,** A firefly bioluminescence ATP assay method for rapid detection and enumeration of brewery micro-organisms, *J. Am. Soc. Brewing Chem.,* 34, 145, 1976.

23. **Sharpe, A. N., Woodrow, M. N., and Jackson, A. K.,** Adenosine triphosphate (ATP) levels in foods contaminated by bacteria, *J. Appl. Bacteriol.,* 33, 758, 1970.

24. **Thore, A., Ansehn, S., Lundin, A., and Bergman, S.,** Detection of bacteriuria by luciferase assay of adenosine triphosphate, *J. Clin. Microbiol.,* 1, 1, 1975.

25. **Hysert, D. W. and Morrison, N. M.,** Studies on ATP, ADP, and AMP concentrations in yeast and beer, *Am. Soc. Brewing Chem.,* 35, 160, 1977.

26. **Miller, L. F., Mabee, M. S., Gress, H. S., and Jangaard, N. O.,** An ATP bioluminescence method for the quantification of viable yeast for fermentor pitching, *Am. Soc. Brewing Chem.,* 36, 59, 1978.

27. **Ryder, D. S., Woods, D. R., Castiau, M., and Masschelein, C. A.,** Glycolytic flux in lager yeast fermentations: application of isotachophoresis and bioluminescence for measurement of cellular intermediates, *J. Am. Soc. Brewing Chem.,* 41, 125, 1983.

28. **Waites, M. J. and Ogden, K.,** The estimation of nisin using ATP bioluminometry, *J. Inst. Brewing,* 93, 30, 1987.

29. **Simpson, W. J., Hammond, J. R. M., Thurston, P. A., and Kyriakides, A. L.,** Brewery process control and the role of "instant" microbiological techniques, *Proc. Eur. Brewing Convention,* 22, 89, 663, 1989.

30. **Avis, J. W.,** The use of ATP bioluminescence for the quality assurance of PET bottled beers, in *Microbiologia Applicata Metodi rapidi ed automatizzati,* Societa Editoriale Farmaceutica, Milan, Italy, 1988, 39.

31. **Jago, P. H., Simpson, W. J., Denyer, S. P., Evans, A. W., Griffiths, M. W., Hammond, J. R. M., Ingram, T. P., Lacey, R. F., Macey, N. W., McCarthy, B. J., Salusbury, T. T., Senior, P. S., Sidorowicz, S., Smither, R., Stanfield, G., and Stanley, P. E.,** An evaluation of the performance of ten commercial luminometers, *J. Biolumin. Chemilumin.,* 3, 131, 1989.

32. **Stanley, P. E.,** Extraction of adenosine triphosphate from microbial and somatic cells, *Methods Enzymol.,* 133, 14, 1986.

33. **Thore, A.,** Bioluminescence assay: extraction of ATP from biological specimens, *LKB Wallac Commun.,* 1979.

34. **Siro, M.-R., Romer, H., and Lovgren, T.,** Continuous flow method for extraction and bioluminescence assay of ATP in baker's yeast, *Eur. J. Appl. Microbiol. Biotechnol.,* 15, 258, 1982.

35. **Harold, F. M., Baarda, J. R., Baron, C., and Abrams, A.,** DIO9 and chlorhexidine: inhibitors of membrane-bound ATPase and of cation transport in *Streptococcus faecalis, Biochem. Biophys. Acta,* 183, 129, 1969.
36. **Simpson, W. J., Hammond, J. R. M., and Miller, R. B.,** Avoparcin and vancomycin: useful antibiotics for the isolation of brewery lactic acid bacteria, *J. Appl. Bacteriol.,* 64, 209, 1988.
37. **Miller, R. and Galston, G.,** Rapid methods for the detection of yeast and Lactobacillus by ATP bioluminescence, *J. Inst. Brewing,* 95, 317, 1989.

Chapter 6

RAPID ENUMERATION OF AIRBORNE MICROORGANISMS BY BIOLUMINESCENT ATP ASSAY

Jillian S. Deans, I. W. Stewart, and T. T. Salusbury

TABLE OF CONTENTS

I.   Introduction..112

II.   Bioluminescent ATP Assay..113

III.   Materials and Methods...113
        A.   Materials...113
                1.   All-Glass Impinger (AGI-30 or Porton Raised Impinger) 113
                2.   Cyclone (Air Centrifuge) 114
                3.   Test Aerosols 114
        B.   Methods...115
                1.   Aerosol Experiments...115
                2.   Conventional Culture Technique 116
                3.   ATP Assay...116
                4.   Cell Viability Experiments..................................116
                5.   Hemocytometer Counting 116

IV.   Results...117

V.   Discussion 121

VI.   Conclusions...126

Acknowledgment..126

References..127

I. INTRODUCTION

Biotechnology is the industrial application of living organisms or biological processes to produce useful products. However, the industry is not new; for thousands of years man has bred improved varieties of crop plants and animals and has exploited the ability of yeast to make alcohol in the form of beer and wine. More recently, over the last 40 years, microorganisms have been used to make antibiotics and other pharmaceuticals. These processes have all used naturally occurring organisms, but productivity of the desired metabolite has been enhanced by methods such as strain selection.

Since 1973, the introduction of the techniques of genetic manipulation or recombinant DNA technology has greatly extended the scope of biotechnology by allowing the transfer of genes between different organisms in a more controlled and predictable manner and often in a way that could not occur naturally through mating and recombination. Genetic manipulation has opened up many new opportunities for industrial application but has also prompted an intense and necessary debate over the anticipation and recognition of potential hazards to health or to the environment and the criteria for practical risk assessment.

An OECD report (1986)[1] has recommended that for good occupational hygiene, tests for the presence of viable process organisms outside areas of primary physical containment should be carried out. Although the microorganisms in the majority of biotechnological applications are unlikely to be pathogenic in themselves, the compounds they produce or express may present a risk to health if the microorganism is able to colonize the body after inhalation. This is in addition to the more obvious problems of direct exposure to these compounds or a complex of cell debris and other products in the air, usually during downstream processing, where a variety of toxic or allergenic risks may be present. In particular, allergenic effects can be induced in some people. Following repeated exposure to a substance, they become sensitized and will then react to extremely low challenge concentrations. The most likely route of occupational exposure is via aerosols generated by breaches of containment during cell growth in bioreactors. Subsequent bioprocessing to separate, concentrate, and purify the desired product may also lead to aerosol formation. Aerobiological samplers are thus vitally important in the detection of containment breaches in bioprocess plants. For the purpose of this report the term ''aerobiological samplers'' is used to describe devices which can collect process microorganisms, which can then be counted by various means.

Most existing aerobiological sampling systems rely on the inertia of airborne particles, either moving under gravity or some external influence, such as an air pump. There are four types of inertial collectors. *Settle plates* and *impactors* rely on airborne organisms settling or impacting onto solid media, while *impingers* and *cyclones* (air centrifuges) impinge organisms into a collection fluid. For impingers, the conventional assay method is to take an aliquot of the collection fluid from the sampler and spread it on to the surface of a nutrient agar plate. All four types of samplers require an associated incubation period, which can be from 12 to 48 h. Only then can microbial colonies be seen and counted, to determine the airborne concentration. For bioprocess safety, techniques which can rapidly count microorganisms in aerobiological samplers are advantageous. They would allow the operators to quickly gain accurate information and thus react rapidly to containment breaches.

The efficiency of collection of the sampling device is an important factor in aerobiological sampling. Each sampler has its associated advantages and disadvantages, but in general, the number of microorganisms recovered is lower than the actual airborne concentration. This is due to loss of viability in the airborne state, ''slippage'' through the sampler, losses on the walls of the samplers, and microorganisms being killed, or being rendered nonculturable, in the sampler. Because of near sonic velocities resulting from sampling, the damage caused by impaction onto solid nutrient media or impingement into collection fluid can be considerable. Deposition in the human respiratory tract is a more gentle process, and the environment is warm and moist, which could encourage recovery. This has important consequences in aerobiological sampling. Colwell et al.[2] and Xu et al.[3] have shown that nonculturable bacteria can retain their ability to transmit plasmids between species and genera. Animal passage has also revealed that pathogenicity may persist in cells classed as "nonviable" by standard culture techniques. Rapid methods of enumeration such as adenosine triphosphate (ATP) assay and the direct epifluorescent technique (DEFT) may provide a means of detecting and counting these viable but nonculturable cells.

II. BIOLUMINESCENT ATP ASSAY

As part of our continuing investigation into the efficiencies of various aerobiological monitoring systems and their application in detecting breaches of containment,[4] we investigated the use of ATP assays for the rapid detection of fugitive microbial emissions.

ATP is an energy-storing substrate that is present in all living cells. The ATP pool in dead cells is rapidly depleted and the concentration of intracellular ATP in healthy cells remains remarkably constant. An assay of ATP in a given sample is thus directly related to the number of living cells present. Microbial ATP can be assayed by a bioluminescent reaction. Microbial assays based on ATP measurement have been used at laboratory scale since 1948, but have suffered from the presence of nonmicrobial ATP, insensitive instruments, and variable reagent quality. These problems have now been largely overcome.[5]

In microbial assays of food and medical samples, nonmicrobial cells must first be treated to release their ATP into solution. Since ATP is a very stable molecule, ATP from nonmicrobial sources, such as food constituents, must then be removed with an ATPase. The remaining viable microbial cells are then treated with a nucleotide-releasing agent, which releases their intracellular ATP into solution. Luciferase, isolated from the firefly *Photoinus pyralis,* and its co-factor luciferin, are then added. The microbial ATP reacts quantitatively with the luciferin-luciferase, producing light, which is measured in relative light units (RLU). The luminescent output is directly proportional to the amount of ATP present, and the ATP concentration in each bacterial species is constant. Since it can be arranged by appropriate choice of collection fluid that aerobiological samples will be free of nonmicrobial cells and free ATP, only the nucleotide-releasing and luciferase steps are required.

In our aerobiological sampling methodology, an impinger and a cyclone were used to sample aerosols of airborne organisms. The organisms present in the collection fluid were then enumerated by conventional colony culture techniques, which required incubation and colony counting after 18 h, as well as by rapid ATP assay, which produced results in less than 30 min.

III. MATERIALS AND METHODS

A. MATERIALS

In order to compare the efficiencies of the samplers, a contained environment was required. This was achieved by the use of a containment cabinet.[4] The cabinet had a 500 l capacity and was fitted with inlet and outlet High Efficiency Particulate Air (HEPA) filters which provided a constant flow of sterile air within the cabinet. Sample ports were arranged along the front in order to accommodate the sampling ports and a glass atomizer.

1. All-Glass Impinger (AGI-30 or Porton Raised Impinger)

Originally described by May and Harper,[6] this was a relatively simple device which consisted of an inlet tube with a 1 mm diameter orifice positioned 30 mm above the bottom of a standard Dreschel bottle. (See Figure 1.) The air flow was controlled by means of a

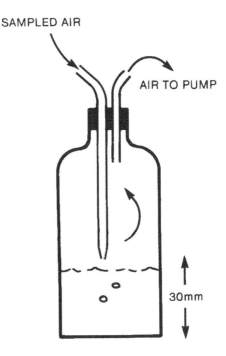

FIGURE 1.   The AGI-30 sampler.

critical pressure drop across the orifice. With a vacuum pressure of greater than 0.5 bar applied by a pump, the air moves through the orifice with sonic velocity which is limiting. The air flow through the sampler was 10 l/min. A rotameter downstream of the pump was used to measure the air flow rate without interfering with sampling. The bottle was filled with phosphate buffer so that the tip of the jet just touched the surface of the liquid and the whole assembly was autoclaved. The sampler was connected via a short length of silicone rubber tubing to a probe inserted into the containment cabinet.

Microorganisms entrained in the sampled air impinged directly into the collection fluid. Results were expressed as colony forming units per l (cfu/l) or RLU/l.

2. Cyclone (Air Centrifuge)

The cyclone sampler designed by Errington and Powell[7] was used (see Figure 2). After sterilization in an autoclave, the cyclone was connected to a 15 mm diameter sampling probe. Air was drawn from the cabinet into the probe and down a short length of silicone rubber tubing into the cyclone at a rate of 750 l/min. Sampled air enters tangentially to the cyclone body, which moves the air in a circular motion. The largest particles impact onto the cyclone wall, while the smaller particles fall out of the air streamlines at the bottom of the cyclone. A metered flow of sterile liquid was injected into the cyclone air inlet to wash the deposited particles into a receiver at the bottom of the cyclone. This fluid was recycled and periodically sampled and assayed. Two sizes of cyclone were used when sampling yeast: the 150 mm model described by Errington and Powell and a 75 mm model of the same linear proportions. Results were expressed as cfu/l or RLU/l.

3. Test Aerosols

The sampling devices were tested with two different spray solutions. *Bacillus globigii* (now known as *B. subtilis* var. *niger*, NCIB 8085) was obtained as freeze-dried cultures from the Torry Research Station, Aberdeen. This bacterium is a Gram positive endospore former, which has been used in numerous studies of impactors and impingers, since it is

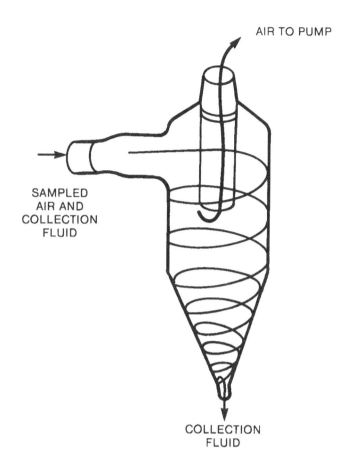

AIR TO PUMP

SAMPLED
AIR AND
COLLECTION
FLUID

COLLECTION
FLUID

FIGURE 2.   The cyclone sampler.

nonpathogenic and forms spores which are relatively resistant to dehydration and physical damage.[8] Stock suspensions of spores were produced in a 2.0 l batch fermenter in a lean nutrient broth medium.[9] Vegetative cells were grown under aerobic conditions at 37°C for 48 h, at which stage the culture had exhausted the carbohydrate supply and sporulated. Heat treatment at 70°C for 30 min killed any remaining vegetative cells. The cells were then pelleted by centrifugation at 10261 g for 15 min and the centrifuged cells washed in sterile phosphate buffer and recentrifuged prior to resuspension and dilution for spraying.

The second spray solution was 5 g of *Saccharomyces cerevisae* in 70 ml 0.1 M phosphate buffer (as a slurry). This was chosen because it is nonpathogenic and widely used in the biotechnology industry. Bakers yeast was purchased in 1 kg blocks (British Fermentation Products, Newport Pagnell) and kept refrigerated until required.

Escherichia coli was used to investigate the possible loss of vegetative cell viability in the AGI-30. Stock suspensions were produced by inoculating 250 ml of the following media, NH_4Cl (5 g/l), KH_2PO_4 (3 g/l), $Na_2SO_4 \cdot 1OH_2O$ (0.5 g/l), Na_2HPO_4 (5.98 g/l), $MgSO_4 \cdot 7H_2O$ (0.2 g/l), glucose (2.5 g/l), trace elements (0.5 g/l), and $FeSO_4 \cdot 6H_2O$ (0.01 g/l), with two *E. coli* colonies from a spread plate. This was then orbitally incubated for 48 h at 37°C, 150 rpm.

B. METHODS

1. Aerosol Experiments

The suspensions were sprayed neat or diluted in pH 7.0 potassium phosphate buffer (0.1 M) into the containment cabinet. A simple glass atomizer was used to produce droplets which rapidly evaporated to leave particles of the desired size. Earlier aerobiologists generated monodisperse aerosols in their studies but we deliberately generated polydisperse aerosols to mimic those produced by leaks from equipment in bioprocessing plants. The aerosol was injected into the cabinet and the system allowed to stabilize before sampling commenced. The droplet size was controlled by varying the liquid/air flow rate, with the resulting particle size distribution (following evaporation of the liquid in the droplet) of the test aerosol determined continuously by optical particle monitors.[4] The aerosol was sampled for a known time, then the collection fluids assayed as described below. Aerosol was continuously generated throughout the sampling period.

2. Conventional Culture Technique

Aliquots (0.1 ml) of collection fluid were taken from the impingers and the cyclone and spread onto a tryptone soya agar plate. These were then incubated for 18 h at 37°C. The resultant colonies were then counted, and the plates incubated for an additional 24 h to enable any damaged organisms further time to recuperate and grow.

3. ATP Assay

Assays were carried out using Lumac reagents and equipment (Lumac® BV Landgraaf, The Netherlands). In samples of collection fluid with a high microbial loading (greater than 1000/ml, usually in the cyclone and in all samples when using yeast), 0.2 ml aliquots of the sampler fluid were assayed directly in the counting chamber of an M2010A Biocounter. 0.2 ml of nucleotide-releasing buffer (NRB) was added, followed by 0.2 ml of Lumit-PM (luciferin-luciferase solution). The luminescent output produced by the released ATP reacting with the Lumit PM was measured as RLU. Standard curves of RLU vs. ATP concentration was constructed for *B. globigii*, *E. coli*, and yeast.

Samples of collection fluid with low microbial loading (usually in the impingers when sampling *B. globigii*) were enriched using 47 mm diameter Whatman® 0.45 μm cellulose filters. The samples were filtered under low vacuum and care was taken not to allow the membrane to dry. The filters were then soaked in 0.5 ml of Lumacult (a peptone-based recovery medium) for 30 min before adding 0.5 ml of NRB. After a contact time of 60 s a 0.2 ml aliquot of the sample was placed in the Biocounter and Lumit was added as before.

4. Cell Viability Experiments

The possibility that aerosolized microorganisms were being inactivated in the sampler was also investigated. A known concentration of cells (determined by hemocytometer count) was placed in an AGI-30. Sterile air was then passed through the sampler at the usual rate to simulate the turbulence of sampling, while aliquots of collection fluid were periodically removed. The samples were assayed using conventional colony culture technique and ATP assay.

5. Hemocytometer Counting

The microbial concentration of the suspensions were determined by hemocytometer counting. Serial dilutions were prepared, selecting a dilution which gave a reasonable count. The suspension was then agitated and a small amount placed in the counting chamber of a Thoma hemocytometer (Hawksley Ltd., Lancing, Sussex). A cover slip was placed over the top of counting chamber to give a depth of 0.1 mm. The number of cells in each of 16 0.05 mm by 0.05 mm squares was counted. The volume observed in each square being known, and the dilution factor taken into account, the number of cells per milliliter of the original suspension was calculated.

TABLE 1
Summary of Results for Comparison of Assay Methods

Date	Assay method	No. of data points	Percentage recovery		Theoretical max. spore conc. (cfu/l)
			Mean	Standard deviation	
12.11.87	CC	6	5.7	1.1	647
	ATP		23.9	3.3	
13.11.87	CC	6	8.0	1.9	647
	ATP		26.9	4.1	
16.11.87	CC	9	5.1	1.8	647
	ATP		26.8	4.1	
19.11.87A	CC	9	7.4	2.2	1042
	ATP		13.0	2.5	
19.11.87B	CC	6	7.8	1.5	2083
	ATP		1.9	1.4	
1.12.87	CC	9	9.7	2.6	182
	ATP		168.6	16.0	
2.12.87	CC	9	7.2	2.8	182
	ATP	8	14.5	9.7	
3.12.87	CC	9	9.0	4.9	182
	ATP		54.5	9.7	

IV. RESULTS

Table 1 illustrates the differences in relative recovery rates obtained in various AGI-30s, at varying spore concentrations, and assaying by conventional colony counting and by ATP assay. As can be seen from Table 1, the normalized percentage recovery of cells is higher when enumerated by ATP assay than by conventional colony counting. Figure 3 shows the average normalized percentage recovery of cells, determined by bioluminescent assay, to that determined by conventional techniques. This appears to indicate that the release of ATP from spores is less efficient at higher concentrations.

Figures 4 and 6 show the apparent loss of viability in the cyclone and AGI-30, respectively, as normal sampling progresses (this is shown by both conventional and rapid techniques in the cyclone).

Because the percentage recovery rates were low in both the AGI-30 and the cyclone, their physical efficiencies were measured using a suspension of salt, whereby the conductivity of the solution was monitored continuously throughout the experiment. The mean and standard deviation values for percentage recovery of the AGI-30 are listed in Table 2. The mean values ranged from 35 to 49%. Figure 5 shows the measurement of physical efficiency of the cyclone. At the end of each test the cyclone was rinsed out and the amount of salt trapped on the walls was assayed. This procedure gave two more values, shown in Figure 5 as "washout", of 82 and 93% recovery. As can be seen, in all cases the physical efficiency was greater than the microbial efficiency.

Figures 7 and 8 show the results for *S. cerevisiae* obtained after sampling with the large and the small cyclones, respectively. The cell concentration (in terms of cells per milliliter of collection fluid) was plotted vs. the time (in minutes) at which a 0.1 ml sample was removed for bioluminescent assay. The figures against each individual curve represent the airborne concentration of cells within the cabinet (in terms of cells per liter of air), calculated using the relevant aerosol injection rates. Both figures show that the concentration of yeast cells in the collection fluid increases in a linear fashion with time. The figures also show that doubling the airborne concentration approximately doubles the concentration of cells in the collection fluids. There appears to be no difference in the efficiencies of the two sizes of cyclone.

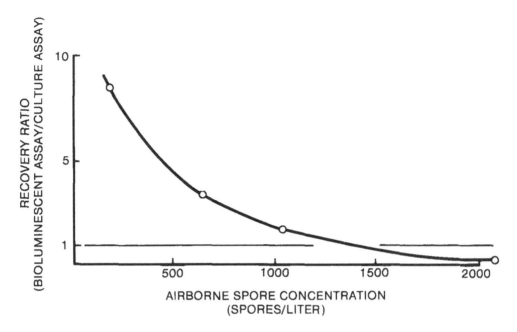

FIGURE 3.   Relative recovery ratio airborne spore concentration in AGI-30.

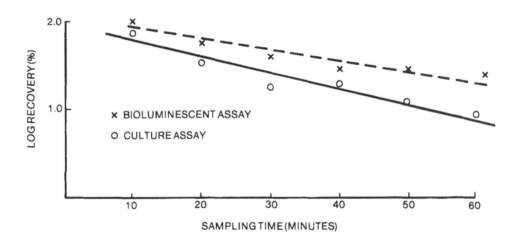

FIGURE 4.   Loss of viability/culturability with time in cyclone sampler.

The results for the AGI-30 are shown in Table 3. The percentage recovery of cells calculated by bioluminescent assay is higher than that calculated by the colony counting technique.

Figures 9, 10, and 11 are calibration curves, log RLU vs. log number of cells for *E. coli*, yeast and *B. globigii*, respectively. They were used to convert RLUs into cfu/ml for use in the cell viability experiments (where the effect of turbulence was investigated). These graphs also give an idea of the amount of ATP per cell.

Figure 12 shows the effect of increasing sampling time on the concentration of *E. coli* in an AGI-30 as determined by both conventional colony counting methods and ATP assay. Both graphs indicate that the concentration remains relatively constant over the 60 min sampling period and the concentration of cells in the impinger corresponds approximately to that determined by hemocytometer counting.

TABLE 2
Percentage Recoveries of Three
AGI-30s Using NaCl

| Date | Sampler | Percentage recovery | |
		(Mean)	(SD)
24.2.88	1	39.9	13.7
	2	49.0	14.2
	3	34.7	13.5

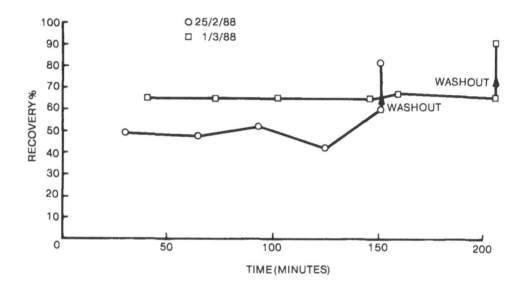

FIGURE 5. Percentage recovery of NaCl in cyclone sampler.

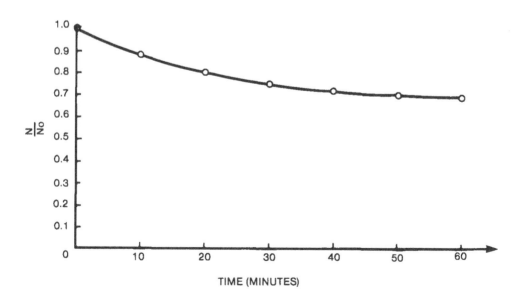

FIGURE 6. Loss of cell viability/culturability in AGI-30s.

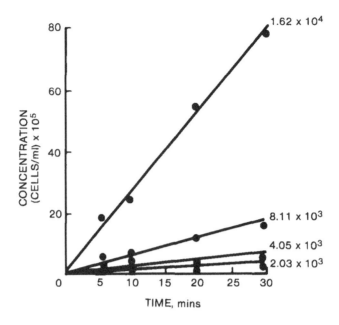

FIGURE 7.   150 mm cyclone sampler cell concentration (determined by ATP assay) vs. sampling times at four aerosol concentrations.

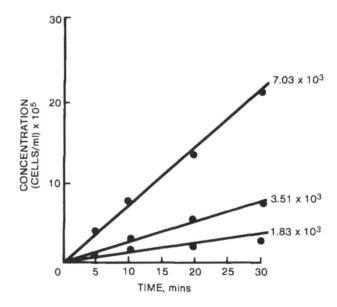

FIGURE 8.   75 mm cyclone sampler cell concentration (determined by ATP assay) vs. sampling times at three aerosol concentrations.

Figure 13 shows the effect of increasing sampling time on the concentration of yeast cells in an AGI-30 as determined by colony counting and ATP assay. The concentration of cells determined by colony counting stays relatively constant and the concentration of cells is of the same order as that determined by hemocytometer counting. The concentration of cells determined by ATP assay shows a slight upwards trend; however, the concentration of the cells is less than that determined by hemocytometer counting.

Figure 14 shows the effect of sampling on *B. globigii* in an AGI-30. The concentration

TABLE 3
Percentage Recovery Rates of Yeast Aerosols in
AGI-30s

Date	No. of data points	ATP assay (mean)	(SD)	Colony counting (mean)	(SD)
20.7.88	9	54.52	21.48	19.0	7.7
21.7.88	9	59.78	30.61	23.8	10.9

With header "Method of enumeration" spanning ATP assay and Colony counting.

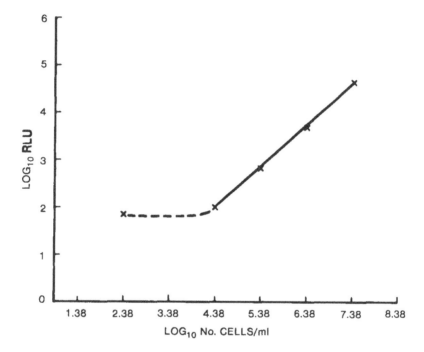

FIGURE 9. *E. coli* calibration curve.

of cells determined by ATP assay is fairly constant while the concentration determined by colony counting shows an overall decrease with time.

V. DISCUSSION

When sampling aerosols of *B. globigii*, both the colony culture technique and ATP assay show that the percentage recovery can apparently decrease with increasing retention in the cyclone. (Figure 4). The physical efficiency (Figure 5, measured by assaying the recovery of sodium chloride aerosols) remains constant with increasing sampling time. It is thus clear that long retention times reduce sample viability, since slippage cannot be a factor here. This is despite the apparent protection afforded by the endospore form, although it is possible that some cells were in the less robust vegetative form, if germination had occurred before aerosolization. Presumably, very short assay times enable viable, damaged cells, which would lose their viability during culturing and incubation, to be enumerated. Since the recovery rates obtained by colony culture technique and ATP assay both decrease, this indicates that viability, not just culturability, is being lost.[10] As the recovery rates obtained when using the ATP assay are higher than those obtained by colony culture technique, some

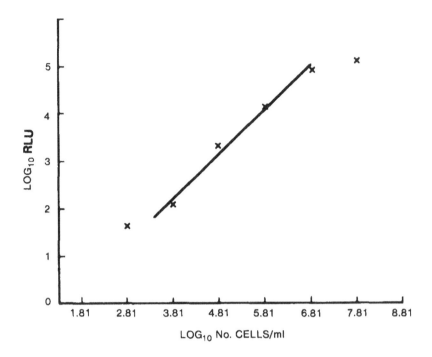

FIGURE 10.   Yeast cake calibration curve.

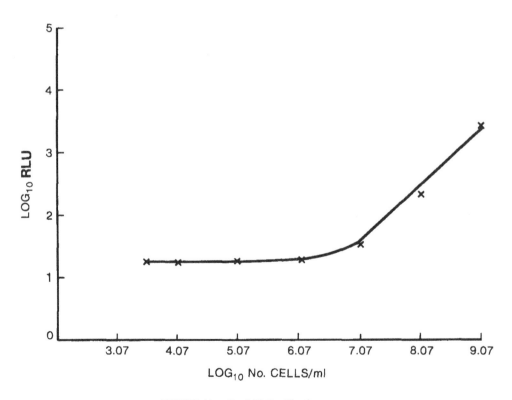

FIGURE 11.   *B. globigii* calibration curve.

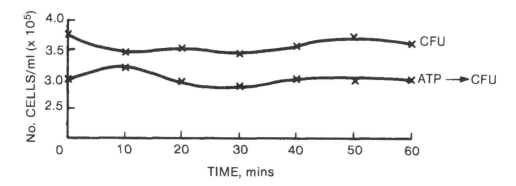

FIGURE 12.   Concentration of *E. coli* vs. time in an AGI-30.

of the cells present may be viable but not culturable. Colwell et al.[2] argue that viable, nonculturable organisms used in bioprocessing plants will present major problems for the industry.

With the AGI-30, determining recovery rates by colony culture technique indicates that there is considerable variability between samplers. These results are by no means atypical, when compared with those obtained in other parts of our experimental program.[11] This has been attributed to loss of viability in the airborne state, losses in the sampling system, variability of individual samples, and microorganisms being killed (or being rendered nonculturable) in the sampler, usually due to impingement at near sonic velocities into the collection fluid.[12] A similar variation is experienced when using the ATP assay, but this technique shows higher recovery rates in all cases but one. This difference in recovery rate appears to be concentration dependent (Figure 3). The reason for this is unclear. It is possible that nucleotide release at the larger spore concentrations was hindered. There may have been insufficient NRB to release all the available ATP or the NRB was physically unable to contact the surfaces of all the spores present.

When sampling yeast, in the cyclones, the recovery rate is constant with time (shown by Figures 7 and 8) in contrast to the results obtained when sampling *B. globigii*. The results also show that for a given sampling time, doubling the concentration of the aerosol feed doubles the concentration of the cells recovered in the collection fluid. This linear relationship could have important consequences in the development of an on-line sampler for the bioprocessing industry, because it is far easier to manipulate instruments and predict data with a linear relationship.

The percentage recovery rates obtained when sampling yeast with the AGI-30 show that higher recovery rates are obtained by bioluminescent assay than by conventional colony counting as was seen when sampling aerosols of *B. globigii*. This indicates that cells containing ATP are being recovered by the sampler but they are not all accounted for by conventional colony counting techniques, so cells are viable but nonculturable.

The calibration curves obtained (Figures 9, 10, and 11) give some idea of the amount of ATP contained in the test organisms used. Clearly, yeast and *E. coli* contain more ATP than the *Bacillus* spores. Elsewhere in our experimental program[11] it was found that the *Bacillus* spores used contained 0.37 fg ATP, while yeast and vegetative *E. coli* contain approximately 1 fg and 1 pg per cell, respectively. This makes the latter two organisms more amenable to enumeration by ATP assay.

Figure 12 shows that the concentration of *E. coli* cells, as determined by colony counting and ATP assay, in an AGI-30 over a 60 min period remains constant. This would indicate that cells are being rendered neither nonviable nor nonculturable.

Figure 13 shows that the cell concentration of the yeast, as determined by colony counting, is constant over the sampling time. However, when determined by ATP assay,

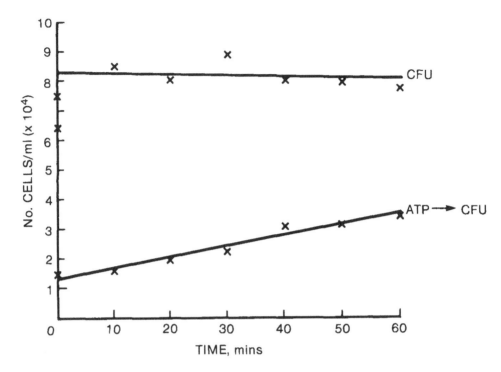

FIGURE 13.   Concentration of yeast vs. time in an AGI-30.

the number of cells is found to increase with time. This increase in ATP could be as a result of cell growth rather than cell division, i.e., the cells are becoming more metabolically active and contain proportionally more ATP. The ATP concentration of fungi changes throughout their growth cycle.[13] The conditions in the impinger can be likened to growth in a bioreactor so it is not surprising that the yeast may become more metabolically active. The concentration of cells determined by ATP assay is lower than that determined by hemocytometer counting. This could be due to the NRB being unable to physically contact all of the yeast cells in the integration period used in the experiment.

Figure 14 shows a decrease in the overall cell concentration of *B. globigii* when the concentration was determined by conventional colony counting. This is despite the apparent protection afforded by the endospore. This has been encountered elsewhere in our experimental program. The cell concentration, when determined by ATP assay, remains relatively constant over the sampling period, so the cells are viable but are nonculturable.

With the exception of *B. globigii*, the microorganisms are not greatly affected by the turbulence of the impinger. Since this is not the case during normal sampling procedure (see Figures 4 and 6) it appears that aerosolization and actual impingement at near sonic velocities considerably weakens the microorganisms. Thus, they are more susceptible to being rendered nonculturable and/or nonviable by lengthy sampling periods in the AGI-30. This organism has a reputation for robustness among aerobiologists; however, it is usually used with much shorter sampling periods.

The maximum assay time using the ATP technique was never more than 35 min. When the microbial loading was high (>1000 spores per ml of collection fluid) the maximum assay time was 2 min. For the conventional culture technique, the minimum time was 12 h, but more typically 18 h was required for the colonies to be clearly visible. Thus, the ATP assay method meets the principle requirement of aerobiological monitoring of containment breaches in bioprocess plants, as results can be obtained rapidly.

Cyclone samplers appear to overcome many of the disadvantages experienced with

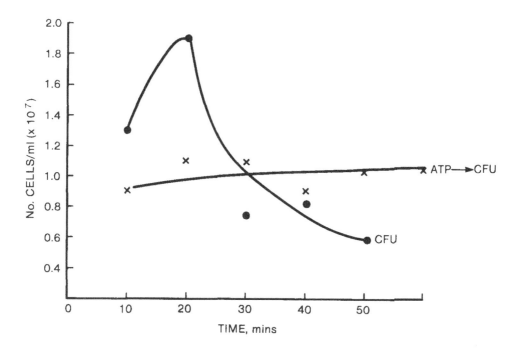

FIGURE 14.   Concentration of *B. globigii* vs. time in an AGI-30.

impingers. Using colony culture techniques, Errington and Powell[7] compared the performance of cyclones with AGI-30 impingers when sampling *B. globigii* spores from an aerosol
chamber. They found that the collection efficiency of the cyclone ranged between 65 and
91%, with the loss in efficiency due to slippage. When sampling an aerosol of *E. coli* with
the cyclone sampler, they reported losses of between 20 to 40% in viability. They also
carried out a field trial where they sampled ambient outdoor air downwind of a manuring
site and compared the performance of small and large cyclones and an impactor. They found
that the air carried 6 cfu per particle, and that the small and large cyclones recorded 0.185
and 0.228 cfu/l of air sampled, while the impactor collected 0.034 cfu/l. This indicates that
aggregated microorganisms in the aerosol were not broken up in the impactors but were in
the cyclone. This suggests that impingers produce results of greater accuracy than may be
obtained with impactors. They concluded that cyclones stress the test organism the least.
When the concentration of airborne microorganisms is expected to be low, a cyclone with
liquid collection is the ideal collection device. Cyclones can sample air at up to 750 l/min,
and can concentrate microorganisms by recirculation of the collection fluid. Thus, they are
amenable to adaptation for continuous operation. However, as the results of this study show,
prolonged sampling can reduce viability.

   Cyclones are thus more suited for sampling operations where large volumes of air must
be handled in a short time period. Examples of operations where release is likely to occur
are liquid sampling from bioreactors and cell harvesting in centrifuges. Cameron et al.[14]
used a cyclone to sample bacterial aerosols generated on the operation of a bioreactor
sampling valve. Sampling time was 15 min.

   As mentioned before, ATP assay meets the principle requirement of aerobiological
monitoring of containment breaches in bioprocess plants. ATP assay is also amenable to
automation and continuous sampling (although not with the particular instrument used in
this study).

   In this study suspensions of *Bacillus* spores and yeast cells were aerosolized. The aerosols
were monitored and collected in samplers. However, several variables which are present in
actual bioprocessing environments have not been covered. These include:

The metabolic state of the cell — Many laboratory-scale studies use stationary-phase cultures from overnight shake flasks. Cultures in vigorous exponential phase from a chemostat would be more representative of bacterial aerosols produced by breaches of containment from bioreactors.

The composition of the aerosol fluid — Aerosols produced from breaches of containment from bioreactors will consist of cells bound in a protective matrix of carbohydrates, protein, and salts, rather than just buffer crystals.

Organism concentration — Since a bioreactor in a bioprocessing plant can contain large concentrations of cells (e.g., 2001 containing 10 cells per ml). A small aerosol particle will thus contain many more organisms than the small particles generated in this study from dilute spore suspensions in buffer.

Despite these omissions, we have gathered enough information to make recommendations as to how bioprocessing environments should be monitored. Four possible strategies have been suggested[10] for routine sampling of biotech process plants with cyclone or AGI-30 samplers.

1. Continuous cyclone sampling, recycling the collection fluid and carrying out ATP assays intermittently
2. Intermittent sampling with an AGI-30, followed by ATP assay
3. Continuous sampling, with one pass of collection fluid through a cyclone, followed by ATP assay
4. Adapting AGI-30s for continuous sampling, with one pass of collection fluid followed by ATP assay.

Our results here show that the first two strategies are likely to encounter problems due to loss of culturability and viability. The third suggestion is possible and promising, while the last is a novel and as yet untried adaptation of existing technology.

VI. CONCLUSIONS

1. ATP assay of aerobiological sampler collection fluids gives a rapid indication of the airborne concentration of microorganisms.
2. ATP assay allows microorganisms which are viable but nonculturable to be enumerated.
3. Certain microorganisms are more suitable for enumeration by ATP assay, e.g., yeast, since they contain more ATP per cell than organisms, in the spore state, such as *B. globigii*. This means that fewer organisms need to be collected to obtain a significant result and a concentration step need not be carried out, thus keeping the assay time to a minimum.
4. Cyclone samplers cause less damage to microorganisms, than AGIs, when used for long-term sampling.
5. AGIs are suitable for short sampling periods. Although simple to construct, one can encounter variabilities in their sampling efficiencies.
6. Aerosolization and impingement of the test aerosol can considerably weaken the microorganisms and render them nonculturable and/or nonviable.

ACKNOWLEDGMENT

This chapter outlines some aspects of our "Risk Assessment in Biotechnology" program, sponsored by the EC DG XII (contract BAP-0110-UK) and the U.K. Health and Safety Executive.

REFERENCES

1. **OECD,** *Recombinant DNA Safety Considerations,* OECD, Paris, 1986.

2. **Colwell, R. R., Brayton, P. R., Grimes, D. J., Roszak, D. B., Hug, S. A., and Palmer, L. M.,** Viable but nonculturable *Vibrio cholerae* and related pathogens in the environment: implications for release of genetically engineered microorganisms, *Bio/Technology,* 3, 817, 1985.

3. **Xu, H.-S., Roberts, N., Singleton, F. L., Attwell, R. W., Grimes, D. J., and Colwell, R. R.,** Survival and viability of nonculturable *E. coli* and *V. cholerae* in the estuarine and marine environment, *Microbiol. Ecol.,* 8, 313, 1982.

4. **Stewart, I. W. and Salusbury, T. T.,** Evaluation and comparison of environmental samplers and particle monitors for bioprocessing plants. Rep. Comm. Eur. Commun., DGXII, Brussels, 1988.

5. **Stanley, C. J., Paris, F., Plumb, A., Webb, A., and Johannssen, A.,** Enzyme amplification — a new technique for enhancing the speed and sensitivity of enzyme immunoassays, *Am. Biotech. Lab.,* May/June, 1985.

6. **May, K. R. and Harper, G. J.,** The efficiency of various liquid impinger samplers in bacterial aerosols, *Br. J. Indust. Med.,* 14, 287, 1957.

7. **Errington, F. P. and Powell, E. O.,** A cyclone separator for aerosol sampling in the field, *J. Hyg. Cambridge,* 387, 1969.

8. **Tyler, M. E. and Shipe, E. L.,** Bacterial aerosol samplers. I. Development and evaluation of the All Glass Impinger, *Appl. Microbiol.,* 7, 337, 1959.

9. **Schaeffer, P., Jonesco, H., Ryter, A., and Balassa, G.,** *Colloque Int. Cent. Natn. Scient.,* Marseille, 124, 553, 1963.

10. **Salusbury, T. T., Deans, J. S., and Stewart, I. W.,** Rapid detection of airborne microorganisms by ATP assay, presented at Proc. Symp. Rapid Microbiology (ATP '88), April 1988, *(Soc. Applied Bacteriology, Tech. Ser.).*

11. **Salusbury, T. T., Deans, J. S., and Stewart, I. W.,** A comparison of six devices used to sample aerosols of *Bacillus subtilis* var *niger* spores. *Proc. 2nd Aerosol Soc. Conf. Aerosols. Their Generation, Behaviour And Application,* 1988.

12. **Cox, C. S.,** *The Aerobiological Pathway of Microorganisms,* John Wiley & Sons, Chichester, England, 1988.

13. **Gaunt, D. M., Trinchi, A. P. J., and Lynch, J. M.,** The determination of fungal biomass using adenosine triphosphate, *Exp. Mycol.,* 9, 174, 1985.

14. **Cameron, R., Hambleton, P., and Melling, J.,** Assessing the microbial integrity of biotechnology equipment, in *Separations For Biotechnology,* Verrall, M. S. and Hudson, M. J., Eds., Ellis Horwood, Chichester, England, 1987, chap. 40.

Chapter 7

THE USE OF ATP MEASUREMENTS IN BIODETERIORATION STUDIES

Brian J. McCarthy

TABLE OF CONTENTS

I.   Introduction 130
      A.   Definitions 130
      B.   Requirement for Rapid Methods 130

II.   Application of ATP Assays to Biodeterioration 132
      A.   Rationale 132
      B.   Case Histories 132

III. Future Possibilities 133
      A.   Improved Enzyme Performance 133
      B.   Bioluminescent Biodeteriogens 133
      C.   Biodeteriogen Interactions 134
      D.   Novel Test Substrates 134
      E.   Novel Test Systems 134

References 134

I. INTRODUCTION

The purpose of this chapter is to provide a brief overview of biodeterioration problems, to present a variety of case histories involving the application of ATP luminescence to biodeterioration studies, and to indicate how recent developments in ATP-related technology may potentially be applied in future investigations of biologically induced spoilage.

A. DEFINITIONS

Hueck,[1] in 1965, provided the widely accepted definition of biodeterioration as "any change in the properties of a material caused by the vital activities of organisms". In a more recent definition, Singleton and Sainsbury[2] stated that biodeterioration may be regarded as deterioration (spoilage) of an object or material as a result of biological — usually microbial — activity. Biodeterioration, in general, refers therefore to a decrease in value of a material or a reduction in the ability of a product to fulfill the function for which it was intended, due to the biological activities of macro- or microorganisms.

Materials subject to biodeterioration range from surface coatings to rubbers and plastics, from fuels and lubricants, to metals and stone.[3] The causative organisms — biodeteriogens — may include bacteria, fungi, algae, yeasts, insects, plants, and animals. Factors affecting the incidence of such deterioration and the subsequent rate and extent of damage are presented in Table 1.

The direct and indirect consequences of biodeterioration may be considerable in terms of financial and material losses. Reported incidents of biodeterioration are characterized by their sporadic nature. Subsequent substrate alterations may range from mild surface discoloration (e.g., surface mold growth on plastic sheeting) to severe structural damage (e.g., termite attack or fungal attack of wooden building supports). In general, the financial impact of biodeterioration processes increases with severity of attack. Therefore, the ability to detect and quantify biologically induced deterioration processes during their initial stages of development will have immediate economic repercussions.

Bioluminescence may be defined as the production of light by living organisms. Examples of luminescent organisms include fungi, dinoflagellates, fish, bacteria, beetles, and, of course, the firefly (*Photinus pyralis*).

The firefly luciferase assay of adenosine triphosphate (ATP) is based on measurement of the light emitted during the luciferase-catalyzed reaction between D-luciferin and $Mg2^+$ — ATP. The requirement for ATP in the firefly luciferase reaction was first demonstrated by McElroy[4] in 1947. The potential analytical usefulness of the firefly reaction for assays of ATP was later confirmed by Strehler and Trotter.[5] Levin et al.[6] used ATP measurements as a sensitive method for the detection of living organisms in natural samples. Holm-Hanson and Booth,[7] in their classic paper, went on to propose that ATP measurements could be used as a measure of total microbial biomass. The central metabolic role of ATP in nearly all cellular energy transactions, the apparent constancy of ATP to organic carbon in a wide range of taxonomically diverse microbes, and the apparent ease of extraction and measurement of ATP from biological samples has stimulated research into the use of ATP as a rapid indicator of biomass.[8] The theoretical principles on which the ATP biomass technique is founded suggest that it is most likely a measure of "protoplasm biomass".

B. REQUIREMENT FOR RAPID METHODS

Laboratory or field investigations of biodeterioration processes may require extended test durations to produce tangible evidence of viability/biodeteriogen activity. Agar plate tests for textiles may take up to 28 days, whereas testing of preservative-treated timber or exposed painted test panels may continue for many months to determine the effects of applied biocides in extending product life. However, many instances exist where rapid methods may

TABLE 1
Factors Affecting the Incidence and Extent of Substrate Biodeterioration

Type of substrate (i.e., natural or synthetic)
Effect of industrial processing (i.e., prior chemical or mechanical damage)
Biodeteriogen distribution
Water activity (i.e., humidity, moisture content)
Temperature and pH
Storage/Packaging
Materials in contact/presence of additives
Exposure to light
Presence of biocides

be used to provide early detection and quantification of biodeterioration.[9] Rapid methods may be taken, in general, to include any procedure which provides results within 24 h. Allsopp and Seal[10] have discussed the various problems that may be associated with the recognition and costing of biodeterioration incidents. In some cases, simple tests may be required merely to confirm the involvement of a biological agency. Stewart[11] has stated that "the detection and enumeration of microorganisms, coupled with the detection and assay of antimicrobial substances, represent key industrial aspects of microbiology. The development of rapid methods to facilitate these assays is of major significance both to provide new and simple protocols and to bring microbial assays into real time, rather than their current retrospective position.[11]

Rapid methods may therefore assist in (1) recognizing biodeterioration problems, (2) physically pinpointing the location of biodeterioration problems, (3) monitoring changes in the properties of a commodity affected by biodeterioration, and (4) providing a fast means of quantifying and confirming the effects of subsequent remedial treatments.

Significant developments have been made in recent years in the application of rapid methods specifically to microbiological biodeterioration. Fung[12] and Colwell[13] have provided general overviews of the various rapid methods available for detection, enumeration, and identification of microorganisms. They have contrasted these novel technologies with traditional culture methods which have long been the mainstay of microbial detection technology. The major drawback of conventional culture methods is again the time required for incubation which may range from 24 h to 7 days, or even longer.

The use of various staining techniques is obviously important in microbial ecology and biodeterioration studies. The acridine orange direct counting procedure[14] is commonly used to obtain total bacterial counts from environmental samples. A modification of this technique has been described[15] which permits the enumeration of cells responding to the presence of a substrate but does not require growth on laboratory media for enumeration. The procedure involves the addition of nalidixic acid and yeast extract to samples, followed by incubation for 6 to 12 h, after which the preparation is fixed and stained. The presence of enlarged cells in the preparation viewed by epifluorescent microscopy indicates utilization of substrate, hence retention of viable cells in total cell populations. Pettipher[16] described the development of a rapid membrane filtration-epifluorescent microscopy technique for the direct enumeration of microbes. Olson et al.[17] have, for example, investigated the disfigurement of external paint films by algae and bacteria using epifluorescence to provide quantitative detection.

Spectrophotometric determination of the hydrolysis of fluorescein diacetate (FDA) has also been suggested as a simple, sensitive, and rapid method of determining microbial activity as a measure of the extent of microbial contamination in/on a substrate. FDA is a nonfluorescent substrate which is hydrolyzed by nonspecific esterases believed to be present in all presumably viable microorganisms.[18] Other rapid methods include the use of radioisotopic labeling,[19] ergosterol assays,[20] electrical methods such as impedance,[21] and calorimetry. DNA probes could offer the potential for the rapid detection and identification of specific microbial biodeteriogens. For example, Festl et al.[22] have developed a DNA probe for *Pseudomonas fluorescens*. Similarly, Turner et al.[23] maintain that novel biosensors could detect biodeterioration by monitoring the formation of a spoilage product or the presence of a specific enzyme which may be released by the causative agent of the deterioration process. However, most of the procedures described above generally require the growth of microbes to produce the amplification needed to bring the cell numbers within the scale of sensitivity of the technique.

II. APPLICATION OF ATP ASSAYS TO BIODETERIORATION

A. RATIONALE

Most biodeterioration investigations require initial detection of biodeteriogen activity, an estimate of the severity of attack (which may or may not be proportional to the biomass present), and confirmation of remedial treatment efficacy. Rapid methods facilitate the provision of results to the industrialist where time is of the essence and may enable the researcher to perform more experiments investigating a greater variety of organism-substrate interactions within the same time constraints.

All living cells, contain ATP (normally associated with proteins).[24] The main advantage of using ATP luminescence lies in its speed and sensitivity.[25] Stanley[27] has provided a concise beginner's guide to rapid microbiology using ATP and luminescence. Additional information has been presented elsewhere.[26-28] ATP luminescence has been used extensively to assess microbial contamination in liquid samples (e.g., urine, process waters, milk, beer, and fruit juices, etc.).[29-30] For appropriate samples assays may be performed in a matter of minutes. Similarly, the technique may be used to detect and measure ATP to below 1 pg (equivalent to 1000 bacterial cells). ATP assays may therefore provide rapid, sensitive, and nonspecies specific estimates of microbial contamination.

The basic protocol involves sample preparation, extraction of somatic ATP (if required), removal of somatic ATP (again, if required), extraction of microbial ATP, and assay via addition of luciferase reagent and measurement of light emission. The provision of novel extractants,[31] a greater awareness of the capabilities of commercially available luminometers,[32] and the ever-increasing availability of standard reagents will stimulate the development of "tailored" protocols for specific applications.

B. CASE HISTORIES

To date, ATP luminescence procedures have been applied to a diverse range of biodeterioration problems. For example Bussey and Tsuji[33] demonstrated that bioluminescence measurements could significantly improve the accuracy, sensitivity, precision, and reliability of the existing visual endpoint determination for the USP sterility test and eliminated the Day-7 transfer/-dilution step required for testing suspension products. Littmann[34] used ATP assays to monitor the microbial quality of hydroxypropyl guar-based fracturing fluids. Jaquess and Hollis[35] evaluated the efficacy of algicides using the luciferase-luciferin enzymatic determination of ATP. Aftring and Taylor[36] used a range of methods, including ATP assays, to investigate microbial fouling in an ocean thermal energy conversion experiment. To assist portability of the technology, Lundblom[37] has described an apparatus, based on ATP luminescence, for registering the presence of bacteria particularly in field conditions.

ATP assays have been applied to biodeterioration studies involving paper,[38] textiles,[9,39,40] packaging materials,[41-43] soils,[43] packaged and processed food,[44] fungal biomass,[45,46] stonework,[47] toxicity determinations,[48] sterility monitoring,[49] contamination monitoring,[50-53] and, of course, biocide efficacy monitoring.[54,55] Some commercial biocides are known to affect the luciferase enzyme directly and particular attention must be paid in protocol development to a residual biocide removal/inactivation step prior to ATP assay.[55] An example of a test

TABLE 2
ATP Luminescence Protocol for Assessing Growth Following Textile Challenge Testing

a. Prepare textile test specimens.
b. Challenge with a standard mixed fungal inoculum.
c. Incubate textile specimens in contact with mineral salt agar at 28°C.
d. Immerse specimens in buffer or extractant.
e. Sonicate or stomach specimens (if required).
f. Remove/inactivate residual biocide.
g. Assay ATP in the resultant liquor.
h. Repeat at intervals during the test period.
i. Use untreated and biocide-treated cotton samples as positive (viability) and negative (biocide inhibition) controls.

Based on BS6085:1981, Methods for determination of the resistance of textiles to microbiological deterioration, British Standards Institution, London.

TABLE 3
Application of ATP Luminescence to Biodeterioration — Advantages and Disadvantages

Advantages	Disadvantages
Speed	Requires specialized reagents
Sensitivity	Enzyme quenching (e.g., biocides, metals, color, etc.)
Estimate of total biomass	Not biodeteriogen specific
Range of equipment available (portable, automated, etc.)	Reagent costs
Quantitative	Lack of standard protocols
Standard reagents available	Presence of background ATP in samples

protocol used for textile biodeterioration studies and based on ATP luminescence is given in Table 2. Hawks and Rowe[56] highlighted the potential importance of airborne microbes in the propagation of spoilage incidents. Salusbury et al.[57] have recently outlined a procedure for rapid detection of airborne microbes by ATP assay which may be of relevance. The ATP luminescence technique offers a range of advantages and disadvantages, some of which are listed in Table 3.

III. FUTURE POSSIBILITIES

How can recent developments in the field of bioluminescence research assist biodeterioration investigations?

A. IMPROVED ENZYME PERFORMANCE

Supplies of r-luciferase, prepared from a prokaryotic recombinant host harboring the *luc* structural gene, are now available from Amgen Biologicals, CA. The use of modern genetic engineering techniques may eventually result in enhanced enzyme stability and bulk availability. Reduced reagent costs may stimulate greater usage of ATP luminescence and related techniques in industrial situations. Similarly, improved resistance to quenching (e.g., pH, heavy metals, pyrophosphate, etc.) and an engineered spectral shift towards the blue region to improve photon detection would be advantageous.[28]

B. BIOLUMINESCENT BIODETERIOGENS

Genetic engineering now allows the introduction of a bioluminescent phenotype into industrially important microorganisms — *in vivo* bioluminescence — which, because subsequent light emission depends on viability, may then be used to monitor *in situ* industrial biocides.[11,58] Using a bioluminescent *E. coli*, Jassim et al.[58] have investigated the phenolic and biguanide biocide classes and have observed significant correlation between concentration dependency of cell death and reduction in observed light emission. Detection of biocide activity could be achieved within a 10 min test period.

The use of firefly luciferase as a reporter gene has been well documented.[59-64] This technology will no doubt greatly facilitate the screening of novel biocidal compounds for industrial usage/commercial exploitation to replace conventional products subject to environmental constraints. Key biodeteriogens could be engineered for specific investigations.

C. BIODETERIOGEN INTERACTIONS

Luminescence assays are generally based on measurements of light alone. Wood[65] has described the color variation between luciferases recovered from the Jamaican click beetle and has suggested their use as paired genetic reporters. Introduction of such reporters into specific microbes could facilitate the observation of microbe-microbe and microbe-biocide interactions in microbial consortia and surface biofilms. The fate of various species could be monitored simultaneously. An immediate area for investigation would be primary microbial settlement during biofouling.

D. NOVEL TEST SUBSTRATES

D-Luciferin derivatives are now commercially available (Novabiochem, Nottingham, U.K.). Bioluminogenic enzyme substrates such as D-luciferin-*O*-phosphate, D-luciferin-*O*-sulfate, D-luciferin-methyl ester, D-luciferin-arginine, etc. allow the ultrasensitive determination of specific enzymes in the femtomole range.[66] Appropriate targets may be synthesized for key enzymes associated with biodeterioration (e.g., cellulases, etc.).

E. NOVEL TEST SYSTEMS

Lundin[67] has described the use of an ATP dipstick (initially for bacteriuria screening) designed to avoid excessive pipetting. The dipstick approach may well have applications in industrial microbiology. New forms of luminescence photobiosensors are being developed.[68,69] Relatively simple camera luminometers have been used to detect luminescent *E. coli*,[70] immobilized bioluminescent enzymes,[71] and biodeterioration,[39] whereas sophisticated charge-coupled devices and photon counting devices offer sensitivity and simultaneous assay[72,73] and image intensifiers offer spacial resolution.[39,74] Finally, biodeteriogen identification procedures may benefit from recent work on bioluminescence in nucleic acid hybridization reactions and the potential development of species specific DNA probes.[75]

REFERENCES

1. **Hueck, H. J.**, The biodeterioration of materials as a part of hylobiology, *Mater. Org.*, 1, 5, 1965.
2. **Singleton, P. and Sainsbury, D.**, *Dictionary of Microbiology and Molecular Biology*, 2nd ed., John Wiley & Sons, Chichester, England, 1987.
3. **Schmidt, O. and Kerner-Gang, W.**, Natural materials, in *Biotechnology 8. Microbial Degradations*, Schonborn, W., Ed., VCH Verlagsgesellschaft, Weinheim, 1986.
4. **McElroy, W. D.**, The energy source for bioluminescence in an isolated system, *Proc. Natl. Acad. Sci. U.S.A.*, 33, 342, 1947.
5. **Strehler, B. L. and Trotter, J. R.**, Firefly luminescence in the study of energy transfer mechanisms. I. Substrate and enzyme determination, *Arch. Biochem. Biophys.*, 40, 28, 1952.
6. **Levin, G. V., Clendenning, J. R., Chappelle, E. W., Heim, A. H., and Roceck, E.**, A rapid method for the detection of microorganisms by ATP assay: its possible application in cancer and virus studies, *Bioscience*, 14, 37, 1964.

7. **Holm-Hanson, O. and Booth, C. R.,** The measurement of ATP in the ocean and its ecological significance, *Limnol. Oceanogr.,* 11, 510, 1966.

8. **Karl, D. M.,** Determination of *in situ* biomass, viability, metabolism and growth, in *Bacteria in Nature, Vol 2. Methods and Special Applications in Bacterial Ecology,* Poindexter, J. S. and Leadbetter, E. R., Eds., Plenum, London, 1986.

9. **McCarthy, B. J.,** Detection and enumeration of micro-organisms on textiles using ATP luminescence, in *ATP Luminescence: Rapid Methods in Microbiology,* Stanley, P. E., McCarthy, B. J., and Smither, R., Eds., Blackwell, Oxford, 1989, 81.

10. **Allsopp, D. and Seal, K.,** *Introduction to Biodeterioration,* Edward Arnold, London, 1986.

11. **Stewart, G. S. A. B.,** *In vivo* bioluminescence: a cellular reporter for research and industry, Pres. Lux Genes Symp., Cambridge, U.K., December 11, 1989.

12. **Fung, D. Y. C.,** Rapid methods and automation in microbiology for biomass estimation, in *Biodeterioration 7,* Houghton, D. R., Smith, R. N., and Eggins, H. O. W., Eds., Elsevier, London, 1988.

13. **Colwell, R. R.,** From counts to clones, in *Changing Perspectives in Applied Microbiology,* SAB Symp. Ser. No. 16, Gutteridge, C. S. and Norris, J. R., Eds., Blackwell, Oxford, 1987, 1.

14. **Hobbie, J. E., Daley, R. J., and Jasper, S.,** Use of nucleopore filters for counting bacteria by fluorescence microscopy, *Appl. Environ. Microbiol.,* 33, 1225, 1977.

15. **Kogure, K., Simidu, U., and Taga, N.,** A tentative direct microscopic method for counting living marine bacteria, *Can. J. Microbiol.,* 25, 415, 1979.

16. **Pettipher, G. L.,** *The Direct Epifluorescence Filter Technique,* Research Studies Press, Letchworth, England, 1983.

17. **Olson, G. J., Iverson, W. P., and Brinkman, F. E.,** Disfigurement of external paint films by algae and bacteria and quantitative detection by epifluorescence microscopy, in *Biodeterioration 6,* Barry, S., Houghton, D. R., Llwellyn, G. C., and O'Rear, C. E., Eds., CAB International, Slough, 1986, 622.

18. **Soderstrom, B. E. and Erland, S.,** Isolation of fluorescein diacetate stained hyphae from soil by micro-manipulation, *Trans. Brit. Mycol. Soc.,* 86, 465, 1986.

19. **Maxwell, S. and Hamilton, W. A.,** Modified radiorespirometric assay for determining the sulphate reduction activity of biofilms on metal surfaces, *J. Microbiol. Methods,* 4, 83, 1986.

20. **Kaspersson, A.,** The role of fungi in deterioration of stored feeds, Rep. Swedish U. Agric. Sci. (Uppsala) No. 31, 1986.

21. **Dickman, M. D.,** The use of impedance monitoring to estimate bioburden, in *Biodeterioration 6,* Barry, S., Houghton, D. R., Llwellyn, G. C., and O'Rear, C. E., Eds., CAB International, Slough, 1986, 419.

22. **Festl, H., Ludwig, W., and Schliefer, K.,** DNA hybridization probe for the *Pseudomonas fluorescens* group, *Appl. Environ. Microbiol.,* 52, 1190, 1986.

23. **Turner, A. P. F., Karube, I., and Wilson, G. S.,** in *Biosensors: Fundamentals and Applications,* Turner, A. P. F., Karube, I., and Wilson, G. S., Eds., Oxford University Press, Oxford, 1987.

24. **Koszegi, T., Berenyi, E., Hazelwood, C. F., Jobst, K., and Kellermayer, M.,** The bulk of ATP is associated to proteins in the living cell — a release kinetics study, *Physiol. Chem. Phys. Med. NMR,* 19, 143, 1987.

25. **Stanley, P. E.,** A review of bioluminescent ATP techniques in rapid micro-biology, *J. Biolumin. Chemilumin.,* 4, 375, 1989.

26. **Schram, E. and Weyens-van Witzenburg, A.,** Improved ATP methodology for bio-mass assays, *J. Biolumin. Chemilumin.,* 4, 390, 1989.

27. **Stanley, P. E.,** A concise beginner's guide to rapid microbiology using adenosine triphosphate (ATP) and luminescence, in *ATP Luminescence: Rapid Methods in Microbiology,* Stanley, P. E., McCarthy, B. J., and Smither, R., eds., Blackwell, Oxford, 1989, 1.

28. **Stanley, P. E. and McCarthy, B. J.,** Reagents and instruments for assays using ATP and luminescence: present needs and future possibilities in rapid microbiology, in *ATP Luminescence: Rapid Methods in Microbiology,* Stanley, P. E., McCarthy, B. J., and Smither, R., Eds., Blackwell, Oxford, 1989, 73.

29. **Chung, Y. C. and Neethling, J. B.,** ATP as a measure of anaerobic sludge activity, *J. Water Pollution Control Fed.,* 60, 107, 1988.

30. **Hanna, B. A.,** Detection of bacteriurea by bioluminescence, *Methods Enzymol.,* 133, 22, 1986.

31. **Simpson, W. J. and Hammond, J. R. M.,** Cold ATP extractants compatible with constant light signal firefly luciferase reagents, in *ATP Luminescence: Rapid Methods in Microbiology,* Stanley, P. E., McCarthy, B. J., and Smither, R., Eds., Blackwell, Oxford, 1989, 45.

32. **Jago, P. H., Simpson, W. J., Denyer, S. P., Evans, A. W., Griffiths, M. W., Hammond, J. R. M., Ingram, T. P., Lacey, R. F., Macey, N. W., McCarthy, B. J., Salusbury, T. T., Senior, P. S., Sidorowicz, S., Smither, R., Stanfield, G., and Stanley, P. E.,** An evaluation of the performance of ten commercial luminometers, *J. Biolumin. Chemilumin.,* 3, 131, 1989.

33. **Bussey, D. M. and Tsuji, K.,** Bioluminescence for USP sterility testing of pharmaceutical suspension products, *Appl. Environ. Microbiol.,* 51, 349, 1986.

34. **Littman, E. S.,** Use of adenosine triphosphate to monitor microbial quality of hydroxypropyl guar based fracturing fluids, in *Biodeterioration 6,* Barry, S., Houghton, D. R., Llwellyn, G. C., and O'Rear, C. E., Eds., CAB International, Slough, 1986, 400.

35. **Jaquess, P. A. and Hollis, C. G.,** Evaluation of algicides by the luciferin-luciferase enzymatic technique, in *Biodeterioration 6,* Barry, S., Houghton, D. R., Llwellyn, G. C., and O'Rear, C. E., Eds., CAB International, Slough, 1986, 631.

36. **Aftring, R. P. and Taylor, B. F.,** Assessment of microbial fouling in an ocean thermal energy conversion experiment, *Appl. Environ. Microbiol.,* 38, 734, 1979.

37. **Lundblom, E.,** Apparatus for registering the presence of bacteria particularly in field conditions, U.S. Patent 4,672,039.

38. **Young-Bandala, L. and Boho, M. J.,** An innovative method for monitoring microbiological deposits in pulp and paper mills, *Tappi J.,* 70, 68, 1987.

39. **McCarthy, B. J.,** Application of bioluminescence techniques to textile biodeterioration, in *Biodeterioration 6,* Barry, S., Houghton, D. R., Llewellyn, G. C., and O'Rear, C. E., Eds., CAB International, Slough, 1986, 394.

40. **McCarthy, B. J.,** Rapid methods for the detection of biodeterioration in textiles, *Int. Biodeter.,* 23, 357, 1987.

41. **Senior, P. S., Tyson, K. D., Parsons, B., White, R., and Wood, G. P.,** Bioluminescent assessment of microbial contamination on plastic packaging materials, in *ATP Luminescence: Rapid Methods in Microbiology,* Stanley, P. E., McCarthy, B. J., and Smither, R., Eds., Blackwell, Oxford, 1989, 137.

42. **Senior, P. S. and McCarthy, B. J.,** Bioluminescent assessment of fungal growth on plastic packaging materials, in *Biodeterioration 7,* Houghton, D. R., Smith, R. N., and Eggins, H. O. W., Eds., Elsevier, London, 1988, 507.

43. **MacLeod, N. H., Chappelle, E. W., and Crawford, A. M.,** ATP assay of terrestrial soils: a test of an exobiological experiment, *Nature,* 223, 267, 1969.

44. **Stannard, C. J. and Gibbs, P. A.,** Rapid microbiology: applications of bioluminescence in the food industry — a review, *J. Biolumin. Chemilumin.,* 1, 3, 1986.

45. **Hendy, N. A. and Gray, P. P.,** Use of ATP as an indicator of biomass concentration in the *Trichoderma viride* fermentation, *Biotechnol. Bioeng.* 21, 153, 1979.

46. **Gaunt, D. M., Trinci, A. P. J., and Lynch, J. M.,** The determination of fungal biomass using adenosine triphosphate, *Exp. Mycol.,* 9, 174, 1985.

47. **Tiano, P., Tomaselli, L., and Orlando, C.,** The ATP-bioluminescence method for a rapid evaluation of the microbial activity in the stone materials of monuments, *J. Biolum. Chemolumin.,* 3, 213, 1989.

48. **Parker, C. E. and Pribyl, E. J.,** Assessment of bacterial ATP response as a measurement of aquatic toxicity, in *Toxicity Screening Procedures Using Bacterial Systems,* Liu, D. and Dutka, B. J., Eds., Marcel Dekker, New York, 1984, 283.

49. **Webster, J. J., Walker, B. G., Ford, S. R., and Leach, F. R.,** Determination of the sterilization effectiveness by measuring bacterial growth in a biological indicator through firefly luciferase determination of ATP, *J. Biolumin. Chemilumin.,* 2, 129, 1988.

50. **Shaw, S.,** Rapid methods for estimating numbers of organisms on surface swabs, in *British Food Manufacturers Industrial Research Association Technical Note No. 4,* Wood, J. M. and Gibbs, P. A., BFMIRA, Leatherhead, 1983.

51. **Blackburn, C de W., Gibbs, P. A., Roller, S. D., and Johal, S.,** Use of ATP in microbial adhesion studies, in *ATP Luminescence: Rapid Methods in Microbiology,* Stanley, P. E., McCarthy, B. J., and Smither, R., Eds., Blackwell, Oxford, 1989, 145.

52. **Ugarova, N. N., Brovko, L. Y., and Lebedeva, O. V.,** Bioluminescence analysis in medicine and biotechnology, *Antibiot. Med. Biotekhnol.,* 31, 141, 1986.

53. **Ugarova, N. N., Brovko, L. Y., Trdatyan, I. Y., and Rainina, E. I.,** Bioluminescent assays in microbiology, *Prikl. Biokhimi. Mikrobiol.,* 23, 14, 1987.

54. **Stanley, P. E.,** Rapid microbiology: the use of luminescence and adenosine triphosphate (ATP) for enumerating and checking effectiveness of biocides: present status and future prospects, in *Biodeterioration 7,* Houghton, D. R., Smith, R. N., and Eggins, H. O. W., Eds., Elsevier, London, 1988, 664.

55. **Denyer, S.,** ATP bioluminescence and biocide assessment: effect of bacteriostatic levels of biocide, in *ATP Luminescence: Rapid Methods in Microbiology,* Stanley, P. E., McCarthy, B. J., and Smither, R., Eds., Blackwell, Oxford, 1989, 189.

56. **Hawks, C. A. and Rowe, W. F.,** Deterioration of hair by airborne microorganisms: implications for museum biological collections, in *Biodeterioration 7,* Houghton, D. R., Smith, R. N., and Eggins, H. O. W., Eds., Elsevier, London, 1988, 461.

57. **Salusbury, T. T., Deans, J. S., and Stewart, I. W.,** Rapid detection of airborne micro-organisms by ATP assay, in *ATP Luminescence: Rapid Methods in Microbiology,* Stanley, P. E., McCarthy, B. J., and Smither, R., Eds., Blackwell, Oxford, 1989, 109.

58. **Jassim, S. A. A., Stewart, G. S. A. B., and Denyer, S. P.,** *In vivo* bioluminescence assay for biocidal agents, Pres. Lux Genes Symp., Cambridge, U.K., December 11, 1989.

59. **Engebrecht, J., Simon, M., and Silverman, M.,** Measuring gene expression with light, *Science,* 227, 1345, 1985.

60. **Gould, S. J. and Subramani, S.,** Firefly luciferase as a tool in molecular and cell biology, *Anal. Biochem.,* 175, 5, 1988.

61. **Palomares, A. J., DeLuca, M., and Helinski, D. R.,** Luciferase as a reporter gene in vegetative and symbiotic *Rhizobium meliloti* and other gram-negative bacteria, *Gene,* 81, 55, 1989.

62. **DeWet, J. R., Wood, K. V., DeLuca, M., Helinski, D. R., and Subramani, S.,** Cloning and expression of the firefly luciferase gene in mammalian cells, in *Bioluminescence and Chemiluminescence: New Perspectives,* Schölmerich, J., Andreesen, R., Kapp, A., Ernst, M., and Woods, W. G., Eds., John Wiley & Sons, Chichester, England, 1987, 369.

63. **DeWet, J. R., Wood, K. V., Helinski, D. R., and DeLuca, M.,** Cloning of firefly luciferase cDNA and the expression of active luciferase in *Escherichia coli, Proc. Natl. Acad. Sci. U.S.A.,* 82, 7870, 1985.

64. **DiLella, A. G., Hope, D. A., Chen, H., Trumbauer, M., Schwartz, R. J., and Smith, R. G.,** Utility of firefly luciferase as a reporter gene for promoter activity in transgenic mice, *Nucl. Acid Res.,* 16, 4159, 1988.

65. **Wood, K. V.,** Luc genes: introduction of colour into bioluminescence assays, Pres. Lux Genes Symp., Cambridge, U.K., December 11, 1989.

66. **Miska, W. and Geiger, R.,** Luciferin derivatives in Bioluminescence-enhanced enzyme immunoassays, *J. Biolumin. Chemilumin.,* 4, 119, 1989.

67. **Lundin, A.,** ATP assays in routine microbiology: from visions to realities in the 1980's, in *ATP Luminescence: Rapid Methods in Microbiology,* Stanley, P. E., McCarthy, B. J., and Smither, R., Eds., Blackwell, Oxford, 1989, 11.

68. **Aizawa, M., Tanaka, M., Ikariyama, Y., and Shinohara, H.,** Luminescence biosensors, *J. Biolumin. Chemilumin.,* 4, 535, 1989.

69. **Blum, L. J., Gautier, S. M., and Coulet, P. R.,** Design of luminescence photobiosensors, *J. Biolumin. Chemilumin.,* 4, 543, 1989.

70. **Wood, K. V. and DeLuca, M.,** Photographic detection of luminescence in *Escherichia coli* containing the gene for firefly luciferase, *Anal. Biochem.,* 161, 501, 1987.

71. **Green, K., Kricka, L. J., Thorpe, G. H. G., and Whitehead, T. P.,** Rapid assays based on immobilized bioluminescent enzymes and photographic detection of light emission, *Talanta,* 31, 173, 1984.

72. **Leaback, D. H. and Hooper, C. E.,** The use of an imaging photon detector in the simultaneous, rapid determination of multiple chemiluminescent and bioluminescent reactions in microtitre volumes, in *Bioluminescence and Chemiluminescence: New Perspectives,* Schölmerich, J., Andreesen, R., Kapp, A., Ernst, M., and Woods, W. G., Eds., John Wiley & Sons, Chichester, England, 1987, 439.

73. **Leaback, D. H., Hooper, C. E., and Pirzad, R.,** The use of imaging luminometers for the simultaneous assay of multiple ATP samples by means of the firefly system, in *ATP Luminescence: Rapid Methods in Microbiology,* Stanley, P.E., McCarthy, B. J., and Smither, R., Eds., Blackwell, Oxford, 1989, 277.

74. **Reynolds, G. T. and Gruner, S.,** A high gain image intensifier-spectroscope system for *in vivo* spectral studies of bioluminescence, *IEEE Trans. Nucl. Sci.,* 22, 404, 1975.

75. **Balaguer, P., Terouanne, B., Boussioux, A-M., and Nicolas, J. C.,** Use of bioluminescence in nucleic acid hybridization reactions, *J. Biolumin. Chemilumin.,* 4, 302, 1989.

Chapter 8

LUMINESCENCE INSTRUMENTATION FOR MICROBIOLOGY

Fritz Berthold

TABLE OF CONTENTS

I.   Introduction 140

II.   Detectors 140

III.   Automatic Reagent Injection 141

IV.   Measuring Chamber and Sample Cuvettes 142

V.   Sensitivity 143

VI.   Quality Control 143

VII.   Examples of Commercial Luminometers 144
        A.   Automatic Luminometer 144
        B.   Semi-Automatic Luminometer 146
        C.   Portable Luminometer 147

VIII.   Companies Supplying Luminometers 147

References 149

I. INTRODUCTION

Instrumentation for the measurement of bioluminescence and chemiluminescence have been subject of many publications.[1-12]

In this chapter we shall adhere to the word "luminometer" for this type of instrument, although other names, like radiometers or photometers, are also used occasionally. Furthermore, in this context the term "luminescence" is meant to include bioluminescence and chemiluminescence only, but not fluorescence, phosphorescence, etc.

Luminometers are produced commercially by several companies. A short list is found at the end of this chapter. The use of noncommercial luminometers today is the exception. Therefore, the discussion will concentrate on design principles for luminometers suited for industrial production.

When associating bio- and chemiluminescence to microbiology, one traditionally thinks of bioluminescent ATP techniques.[13] In recent years, however, immunoassay[14-16] and DNA probe methods[17,19] are also employing bio- and chemiluminescent labels. These assay technologies hold the promise to add specificity to sensitivity in rapid microbiology. In consequence, the focus of luminometer design was not limited to ATP technology, but included application in luminescence immunoassay and DNA probe assays.[12] In fact, these latter two applications of luminescence have had a much more dynamic development in recent years than ATP techniques, leading to a new generation of luminometers suited for clinical use.[12] These luminometers are characterized by features like user friendliness, increased reliability, a higher degree of automation inlcuding sample preparation, automatic quality control of instrumentation and reagents, and advanced software reporting final results. It may be expected that these features will also be welcome in rapid microbiology using luminescence. Consequently, a user might prefer a luminometer suitable for all present and future types of luminescence applications in microbiology. The following discussion concentrates on luminometers suitable for ATP assays, luminescence immunoassays, and DNA probe assays with luminogenic labels. General and basic considerations in luminometer design will be presented, hopefully outlasting specifications of individual luminometers.

Those key components which are specific to luminometers, and not familiar from other types of analytical instrumentation will be discussed in some detail: the detector, automatic reagent injectors, and the measuring chamber.

Three commercial luminometer models will then be presented as examples of contemporary luminometer design.

II. DETECTORS

The detectors used in commercial instruments are, almost exclusively, photomultipliers. A few instruments using photodiodes have been produced in the past, but could not meet the requirements of high sensitivity. Recently developed solid-state detectors like avalanche diodes might find use in future luminometers, if present limitations from high background and very limited sensitive area can be overcome.

An imaging photon counter has been applied to measure bioluminescence.[7] The application of an imaging luminometer using charge-coupled devices for immunoassays has also been reported.[9]

Such imaging systems have potentially attractive features, like the possibility to measure an entire microplate simultaneously. But further developments are required before they could be applied routinely in rapid microbiology. This includes the challenge to bring the cost of imaging systems into a range comparable with conventional luminometers.

The photomultipliers used in commercial luminometers have bialkali or multialkali type photocathodes. Their spectral range extends from about 390 to 630 nm. Their quantum efficiency shows a broad maximum of 20 to 28% in the wavelength range of 400 to 450 nm. (Quantum efficiency is the ratio of photoelectrons released to the number of photons impinging on the photocathode.) This matches perfectly with the emission wavelength of luminol or acridinium ester of about 420 to 430 nm. For ATP bioluminescence, at 560 nm, however, the quantum efficiency for bialkali type photomultipliers is reduced to about 3 to 6%, according to the tube model and individual efficiency variations. Photomultipliers with better "red" sensitivity are available, but this advantage is more than offset by much higher background, so that bialkali photomultipliers appear to offer the best compromise.

Photomultipliers can be operated by measuring their anode current, or by single-photon counting. The latter method leads to higher sensitivity.[3,20-23] It is not possible to quote a fixed factor for the gain in sensitivity obtained by applying photon counting instead of current measurement, because this depends on photomultiplier parameters like the noise spectrum, and the contribution from dynode noise, which are different from tube to tube. For some indications of the advantages of photon counting see the section on sensitivity.

Besides superior sensitivity, photon counting leads to better long-term stability, compared to current measurement.[3] This is due to the fact that photon counters are operated in a plateau region, so that small gain drifts in the dynodes and electronic circuits have only negligible influence on sensitivity. As a practical consequence, a luminometer with a photon counter operates with a factory-set high voltage which, except in the case of an instrument failure, does not have to be readjusted by the user.

The noise count rate of a photomultiplier tube is strongly temperature dependent. While typical background count rates of a 1-in-cathode photomultiplier tube at 25°C are in the range of 50 to 250 counts per second these values can be reduced by a factor of about four by cooling to 6°C.

Instrument background, even without detector cooling, is normally at least one order of magnitude smaller than reagent background, and therefore, not the limiting factor for the system sensitivity. This might change if reagent quality should be substantially improved in the future.

The dynamic range of a photon counter luminometer is about 4 decades. Beyond about 4×10^6 counts per second nonlinearities occur, due to dead-time losses.

Relatively high light levels might occur during the measurement of enzyme labels with chemiluminescent readout, such as enhanced luminescence,[16,24] or dioxetane.[25,26] A good method to shift the dynamic range of a photon counter to higher light intensities is the insertion of an optical filter between sample and photocathode.[12]

III. AUTOMATIC REAGENT INJECTION

For routine operation, automatic reagent injection is mandatory, both for single-sample and fully automatic luminometers.

Chemiluminescent labels like cyclic hydrazides (luminol) or acridinium esters exhibit fast kinetics. More than 95% of the light is emitted in about 2 s so that measuring times of 2 s per sample are usually sufficient, allowing high sample throughput. These widely used labels require, because of their fast kinetics, reagent injection with the sample in the measuring position, in front of the photocathode. In the case of acridinium ester, the standard procedure is an acid peroxide injection, followed by a second injection of a weak NaOH solution after 1 to 5 s. Therefore, two reagent injectors are required.

Enzyme labels with luminescence readout, like enhanced luminescence[16] or 1.2-dioxetane,[25,26] show very slow kinetics. Reagent injection in the measuring position is therefore not required. Sometimes it is not even practical, as in the case of 1,2-dioxetanes,[25] where the time to reach stable emission is about 15 min after reagent addition.

ATP/firefly luminescence is frequently claimed to provide a virtually constant light signal with quality commercial reagents. Therefore, it would appear possible to inject luciferin/luciferase outside the luminometer, manually or automatically, followed by insertion of the sample into the luminometer and subsequent light measurement.

However, it has been shown that certain surfactants and similar compounds which are routinely used in commercial extraction agents for microbial ATP, change the kinetics of light emission.[27] This could lead to incorrect results if the user is not alerted to the problem.[13] It is, therefore, preferable to inject luciferin/luciferase while the sample is in the measuring position, so that the kinetics may be monitored from the beginning of the reaction. An example of automatic kinetics monitoring will be presented, together with the description of the automatic luminometer LB 953.

The reagent injectors should be designed to ensure that the reagents will be in contact with totally inert material, e.g., Teflon®, only. This prevents long-term damage to the injectors by aggressive reagents, and also avoids erroneous measurements due to deterioration of the activator reagents as a consequence of chemical interaction between reagents and injector materials.

The injector should ensure prompt mixing of activator reagents and samples. This can be achieved by rather intense injection using a liquid jet.

In ATP measurements, bacterial contamination of the injection system might lead to incorrect results. This can be avoided by filling the injection system with an antimicrobic solution as long as the luminometer is in standby condition.

IV. MEASURING CHAMBER AND SAMPLE CUVETTES

The type of sample cuvette to be used is an important parameter in the design of a luminometer, especially its measuring chamber.

It will be advantageous to the user if cuvettes are of a type generally available, and not specific to one instrument only. An added benefit is the possibility to accommodate more than one type of sample cuvette.

One might distinguish single-tube luminometers — which might be both manual or automatic — and microplate luminometers, which are always automatic.

The most popular cuvette dimensions for single-tube luminometers are 12 × 47, 12 × 50, 12 × 55, and 12 × 75 mm. Cuvettes with low background should be selected. Instrument manufacturers are usually able to supply suitable cuvettes, or to give recommendations.

To obtain optimal sensitivity, the design of the measuring chamber should ensure that the highest possible fraction of photons produced in the reaction reach the sensitive area of the detector.[5] However, for many routine applications other aspects may even be more important: (1) convenient loading and unloading of samples, (2) easy cleaning in case of reagent spills, (3) use of noncorrosive materials, and (4) compatibility with automated sample transport mechanisms.

Luminometers for microplates can not use the standard transparent plates because of optical "crosstalk" between adjacent wells, Therefore, they require plates manufactured from nontransparent material. The use of reflecting materials enhances sensitivity.

Instead of complete microplates, linear strips of 8 or 12 wells, in the microplate dimensional pattern of 9 mm from center to center, might also be employed.

The microplate format allows a high degree of automation and sample throughput, not only for measurement, but also for sample preparation. Most manufacturers place the photomultiplier above the microplate, with the photocathode facing downward, in order to measure the light which is emitted upwards from the wells. An x-y scanning table presents each well successively to the detector. Great care has to be taken to ensure that optical crosstalk between adjacent wells falls below an acceptable level. This requires some means of light collimation. Because light collection in a microplate luminometer is less efficient than for single-tube luminometers, the sensitivity is generally lower.[10]

V. SENSITIVITY

The sensitivity of a luminometer depends on several parameters. The most important ones are the quantum efficiency and the background noise of the photomultiplier tube, and the light collection efficiency of the measuring chamber.

It has been pointed out that chemical or reagent background, meaning the result obtained with all reagents added, but with zero analyte concentration, is often considerably higher than instrument background alone.[10] Therefore, assay sensitivity frequently is not determined by instrument sensitivity, but by reagent properties.

Technical brochures of commercial luminometers usually specify sensitivity. Since photon standards with known emission rate of photons, and the same emission spectrum as the firefly luciferase reaction, are not available, the sensitivity of instrumentation is frequently stated as the minimum amount of ATP which can be detected.

It has been observed that details of how these values were determined are often lacking.[10] This situation has prompted an assessment of the sensitivity for measuring ATP comparing ten luminometers,[10] stating the criteria used to determine detection limits.[28] One of the observations was that among all luminometers tested those using photomultiplier tubes showed the highest sensitivity. A camera luminometer using instant film was found to be 5000 times less sensitive than the best photomultiplier instrument.

The most sensitive luminometer in the test allowed detection of about 0.2 pg ATP using presently available reagents. This value could be improved to about 0.1 pg ATP if background-free reagents were available.

Assuming that 1000 "average" bacterial cells contain 1 pg ATP,[13] one would arrive at a theoretical detection limit of 200 or 100, respectively, microbial cells. Because of quenching effects in biological samples, the practical detection limit is about 1000 bacterial cells.[13]

Even higher sensitivity for microbe detection is achieved with DNA probe technology. The detection limit for chlamydia trachomatis using Gen-Probe's acridinium ester-labeled probes and their LEADER I luminometer, manufactured by Berthold, is better than ten microorganisms.[18]

Analyzing the results by Jago et al.,[10] one must conclude that microplate luminometers do not reach the sensitivity of the best single-tube instruments. The luminometers found to be most sensitive were the (single-tube) luminometers Lumac M 2010 A and Turner 20 TD with detection limits, respectively of 0.20 and 0.27 pg ATP per assay, based on the measurement of one sample and using commercial reagents.

Two microplate readers, from Amersham and Dynatech, were reported to have detection limits of 1.79 pg ATP under essentially the same conditions.

This considerable difference in sensitivity might be attributed to the difficulty of achieving good light collection efficiency for microplates while, at the same time, minimizing crosstalk between adjacent wells. The use of discrete tubes certainly makes it easier to design measuring chambers with high optical efficiency and practically no crosstalk.

Yet another comparison of instrument sensitivity was made in connection with the application of the firefly luciferase as a reporter gene.[26] With an LKB 1250 luminometer 0.30 pg of luciferase was detected, while the A.L.L. Monolight 2001, produced by Berthold and using photon counting, detected 0.01 pg luciferase.

VI. QUALITY CONTROL

For the absolute calibration of a luminometer, a photon standard with a precisely known rate of photon emission, at the same time matching the emission spectrum of the luminescence system, e.g., ATP/bioluminescence, would be required. However, photon standards meeting these requirements are presently not available.

Most luminescence measurements do not reqire an absolute calibration of the luminometer, since unknown samples are usually evaluated by comparison to standards.

It would, therefore, be sufficient to check the long-term stability of a luminometer with a stable photon source, albeit of not precisely known emission rate and spectrum. Such photon check sources are available.

Weak radioactive emitters suitably combined with liquid or plastic scintillators are useful check sources. Quenched tritium standards are a good choice because of their single-photon emission.[30]

Checking luminometer performance with a photon check source is, however, incomplete since it checks the light detector function only. But, especially in a luminometer using a photon counter, the light detector is a very stable component. It is more important to check the quality (and even the presence) of starter reagents, and the function of automatic injectors. This has led to performance checks with liquid calibrators.

A performance check routine monitors the response of a luminometer to a standardized amount of label or liquid calibrator, thus including the entire process of luminescence initiation through injection of starter reagents.

This response is compared to a target value. Target values are established, in a nonroutine procedure, as the mean value of many replicate readings for the same liquid calibrators.

The luminometers LB 953 and LB 9501 described later have programs for automatic performance checks. The deviation of the actual response from the target value is calculated and printed out. Any deviation exceeding a user-defined tolerance range is automatically flagged.

VII. EXAMPLES OF COMMERCIAL LUMINOMETERS

Commercially available luminometers have been described previously.[1-6,8,10-12] But such reports are soon outdated, given that analytical instrumentation is changing rapidly. Many of the luminometers described earlier are no longer available, or have been replaced by new models.

The luminometers described here are recent introductions, and have been chosen as examples only. For the reasons mentioned, it has not been attempted to compile a complete list of all commercially available luminometers.

The discussion of the automatic luminometer LB 953 is intended to demonstrate how instrument design follows the measuring and sample preparation protocol as closely as possible, and how hard- and software design might help to overcome some problems associated with the chemistry, or at least alert the user to such problems.

A. AUTOMATIC LUMINOMETER

The automatic luminometer LB 953 from Berthold (Figure 1) is designed for universal use in ATP-measurement, immunoassay, DNA probe assay, and cellular chemiluminescence. Software is available for all fields of application, either using a built-in microprocessor or an external computer.

An automatic sample changer for 150 samples allows single or repeated runs for the same samples. Any tube from 12 × 47 to 12 × 75 mm can be accommodated.

Three reagent injectors are positioned at (1) the measuring position (position zero), (2) at one sample position before (position minus 1), and (3) at two sample positions before the measuring position (position minus 2). The injectors are under computer control, and each one can be triggered individually.

The sample compartment may be temperature controlled from ambient to 45°C, thereby functioning also as an incubation area.

The standard detector is a photon counter covering a spectral range from 390 to 620

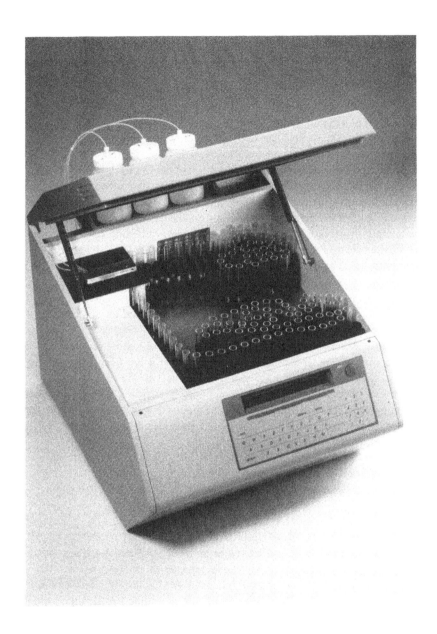

FIGURE 1. Automatic luminometer LB 953 (Berthold).

nm. The standard photomultiplier can easily be replaced by other types with different spectral response. To reduce background, the photomultiplier is cooled to 6°C.

The LB 953 uses an integrated 16-bit microprocessor system, providing sufficient capacity for all operating and evaluation software. Host computers may be connected through a bidirectional RS 232 C interface, which also allows complete operation of the instrument under external computer control.

The following describes the protocol of a completely automated ATP-assay with the LB 953 luminometer, according to the procedure described by Stanley.[13]

Extraction of somatic ATP and removal of free and somatic ATP — In commercial reagents, the extraction agent for somatic ATP is normally mixed with an ATPase to remove somatic and free ATP. This reagent mixture is automatically injected at position minus 2, and the samples are then returned to the sample compartment. During a preset time, typically 10 to 30 min, the samples are incubated with the extraction agent. Temperature control is possible.

Extraction of microbial ATP — As soon as the preset incubation time is over, the sample transport is started automatically, and the microbial extraction agent is injected automatically into the sample tube at position minus 2. Normally a short incubation time of less than 10 s is sufficient, so that the injection of luciferin/luciferase at position zero may follow immediately. If much longer incubation times are desired, the samples may be automatically transported back into the sample compartment, after injection of the microbial extraction agent only, and they are measured after the end of a second incubation time.

Addition of luciferin/luciferase and measurement — Either immediately after injection of the microbial extraction agent, of following a second incubation period as mentioned above, the luciferin/luciferase is injected in position zero, initiating light emission. Optionally, an automatic background measurement may be performed immediately before injection of luciferin/luciferase. The background value thus obtained can be automatically subtracted from the result of the subsequent measurement. The beginning of pulse counting can be delayed with respect to the reagent injection in steps of 0.1 s. In some applications, improved precision is obtained if integration begins only after reaching the phase of practically stable light emission.

It is frequently asserted that the light output for purified commercial luciferin/luciferase is practically constant over an extended time period. But it has been observed that certain surfactants change the kinetics of light emission.[27] The LB 953 allows monitoring of the light emission kinetics automatically. This is achieved by dividing the total preset measuring time into 20 equal time intervals, and calculating a time-intensity histogram. Irregular kinetics are automatically detected. A kinetics printout is possible; this feature may be limited to plotting abnormal kinetics only.

Another possible way of monitoring irregular kinetics is to remeasure all samples automatically after another preset waiting time without further reagent injection, and to compare the results with those obtained immediately after luciferin/luciferase injection.

B. SEMI-AUTOMATIC LUMINOMETER

The luminometer LB 9501 from Berthold (Figure 2) is microprocessor based and uses a photon counter like the LB 953, operated at ambient temperature. It has two automatic reagent injectors. For ATP-measurement, one injector might be used for the microbial extraction agent, (100 or 300 µl), and the other one for luciferin/luciferase addition (50 or 100 µl). Special care has been taken to minimize dead volume for the luciferase injectors, in order to reduce consumption of expensive reagent during injector priming.

Both hard- and software have been designed to make the LB 9501 universally suitable for ATP-assays, luminescence immunoassay, DNA probe assays, etc. It may be noteworthy that it is widely used for reporter gene studies using the luciferase gene,[29] requiring highest sensitivity for luciferase detection.

The LB 9501 has a built-in printer, and an LS 232 C interface for connection of external computers.

Available software includes

1. Calculation of count rates for single and replicate samples including coefficients of variation, with automatic background compensation
2. "Cut-off" protocols classifying results into positive (contaminated) and negative (non-contaminated) samples.[11] This software is used both in ATP and DNA probe assays
3. Complete immunoassay evaluation, for competitive (LIA) and immunochemilumetric assays (ICMA)

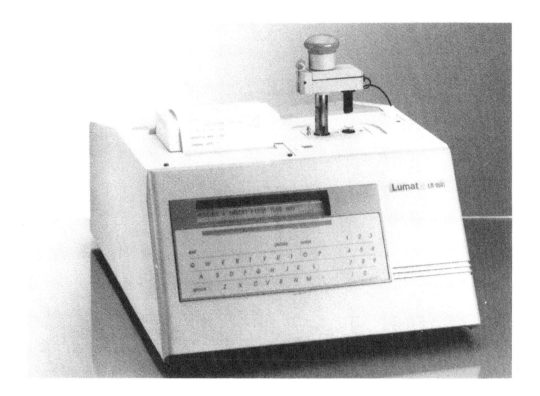

FIGURE 2. Semi-automatic luminometer Lumat LB 9501 (Berthold).

C. PORTABLE LUMINOMETER

The Lumac Biocounter M 1500 P (Figure 3) is a portable instrument, powered by built-in rechargeable batteries or mains. Its preferred use is in rapid microbial testing based on ATP measurement. The instrument is part of a purpose-built case, providing room for reagents, pipettes, cuvettes etc. A result printer is also included. Reagent addition has to be performed manually.

The detector is a photon counter; therefore, the luminometer is expected to be equally sensitive as the Lumac 2010 A, mentioned in the section discussing sensitivity. Cuvette format is 12×47 mm.

VIII. COMPANIES SUPPLYING LUMINOMETERS

At present more than twenty different models of luminometers are on the market. Photomultipliers are the dominant detectors by far. The more recent introductions favor photon counting over current measurement. A camera luminometer is manufactured by Dynatech.

The majority of luminometers sold presently are used for clinical luminescence immunoassays.[12] Other important fields of application are bioluminescence,[13] molecular biology,[28] and cellular chemiluminescence,[31] The vast field of luminometer applications, using commercial or noncommercial instruments, may be appreciated by studying the proceedings of one of the most recent major luminescence symposia.[32]

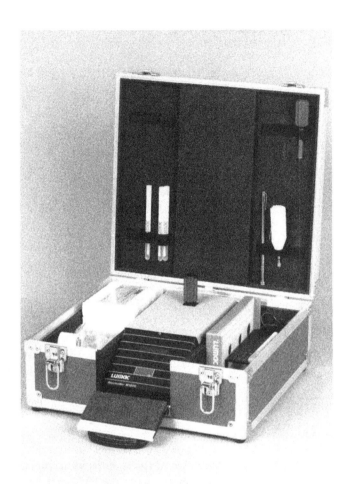

FIGURE 3.   Portable luminometer Lumac Biocounter M 1500 P (Lumac).

Manufacturers include:

Amersham International PLC, Amersham, U.K.
Berthold Laboratorium, Wildbad, Germany
Bio-Orbit, Turku, Finland
Dynatech Laboratories, Billingshurst, U.K.
Flow Laboratories, Rickmansworth, U.K.
Foss Electric, Hillerod, Denmark
Hamilton Bonaduz AG, Bonaduz, Switzerland
Los Alamos Diagnostics Inc., Los Alamos, New Mexico, U.S.
Lumac BV, Landgraaf, The Netherlands
Turner Designs, Mountain View, CA, U.S.

REFERENCES

1. **Stanley, P. E.**, Instrumentation, in *Clinical and Biochemical Luminescence*, Kricka, L. J. and Carter, T. J. N., Eds., Marcel Dekker, New York, 1982, 219.
2. **Stanley, P. E.**, Instrumentation for luminescence methods of analysis, *Trends Anal. Chem.*, 2, 248, 1983.
3. **Berthold, F.**, Measurement of bio- and chemiluminescence — liquid scintillation counters vs. dedicated luminometers, in *Advances in Scintillation Counting*, McQuarric, S. A., Ediss, C., and Wiebe, L. I., Eds., Faculty of Pharmacy and Pharmaceutical Sciences, University of Alberta, Edmonton, Canada, 1983, 230.
4. **Wood, W. G., Gadow, A., and Strasburger, C. J.**, Comparison of two semiautomatic luminometers suitable for routine luminescence immunoassays, in *Analytical Applications of Bioluminescence and Chemiluminescence*, Kricka, L. J., Stanley, P. E., Thorpe, G. H. G., and Whitehead, T. P., Eds., Academic Press, London, 1984, 465.
5. **Turner, G. K.**, Measurement of light from chemical or biochemical reactions, in *Bioluminescence and Chemiluminescence: Instruments and Application*, Vol. 1, Van Dyke, K., CRC Press Boca Raton, FL, 1985, 43.
6. **Berthold, F. and Maly, F. E.**, Instrumentation for cellular chemiluminescence, in *Cellular Chemiluminescence*, Vol. 1, Van Dyke, K. and Castranova, V., CRC Press, Boca Raton, FL, 1987, 49.
7. **Leaback, D. H. and Hooper, C. E.**, The use of an imaging photon detector in the simultaneous, rapid, determination of multiple, chemiluminescent and bioluminescent reactions in microlitre volumes, in *Bioluminescence and Chemiluminescence: New Perspectives*, Schölmerich, J., Andreesen, R., Kapp, A., Ernst, M., and Woods, W. G., Eds., John Wiley & Sons, Chichester, England, 1987, 439.
8. **Leyssens, H., Camenisch, J. L., Hardmeier, B., Saba, H., Schram, E., and Roosens, H.**, Single-detector, multi-head luminometer with integrated computer, in *Bioluminescence and Chemiluminescence: New Perspectives*, Schölmerich, J., Andreesen, R., Kapp, A., Ernst, M., and Woods, W. G., Eds., John Wiley & Sons, Chichester, England, 1987, 587.
9. **Leaback, D. H. and Haggart, R.**, The use of a CCD imaging luminometer in the quantitation of luminogenic immunoassays, in *Journal of Bioluminescence and Chemiluminescence*, Vol. 4, Pazzagli, M., Cadenas, E., Kricka, L. J., Roda, A., and Stanley, P. E., John Wiley & Sons, Chichester, England, 1989, 512.
10. **Jago, P. H., Simpson, W. J., Denyer, S. P., Evans, A. W., Griffiths, M. W., Hammond, J. R. M., Ingram, T. P., Lacey, R. F., Macey, N. W., McCarthy, B. J., Salusbury, T. T., Senior, P. S., Sidorowicz, S., Smither, R., Stanfield, G., and Stanley, P. E.**, An evaluation of the performance of ten commercial luminometers, in *Journal of Bioluminescence and Chemiluminescence*, John Wiley & Sons, Chichester, England, 1989, 131.
11. **Berthold, F.**, Luminometers for ATP Assays, immunoassays and DNA probe assays, in *ATP Luminescence: Rapid Methods in Microbiology*, Society for Applied Bacteriology, Tech. Ser., Vol. 26, Stanley, P. E., McCarthy, B. J., and Smither, R., Blackwell, Oxford, in press.
12. **Berthold, F.**, *Instrumentation for Chemiluminescence Immunoassays*, Van Dyke, K., Ed., CRC Press, Boca Raton, FL, 1990, 11.
13. **Stanley, P. E.**, A review of bioluminescent ATP-techniques in rapid microbiology, in *Bioluminescence and Chemiluminescence: Studies and Applications in Biology and Medicine*, Pazzagli, M., Cadenas, E., Kricka, L. J., Roda, A., and Stanley, P. E., John Wiley & Sons, Chichester, England, 1989, 375.
14. **McCapra, F. and Beheshti, I.**, Selected chemical reactions that produce light, in *Bioluminescence and Chemiluminescence*, Vol. 1, Van Dyke, K., CRC Press, Boca Raton, FL 1985, 9.
15. **Weeks, I., Sturgess, M., Brown, R. C., and Woodhead, J. S.**, Immunoassays using acridinium esters, in *Methods Enzymol.*, 133, 336, 1986.
16. **Thorpe, G. H. G. and Kricka, L. J.**, Enhanced chemiluminescence for horseradish peroxidase: characteristics and application, in *Bioluminescence and Chemiluminescence: New Perspectives*, Schölmerich, J., Andreesen, R., Kapp, A., Ernst, M., and Woods, W. G., Eds., John Wiley & Sons, Chichester, England, 1987, 199.
17. **Granato, P. A. and Franz Roefaro, M.**, Evaluation of a prototype DNA probe test for the noncultural diagnosis of gonorrhea, *J. Clin. Microbiol.*, Vol. 27, No. 4, 632, April 1989.
18. **Bruni, J., Enns, B., Fletcher, R., Lawrence, T., Roeder, P., and Trainor, D.**, Clinical studies summary report, the Gen-Probe Pace ® assay system for Chlamydia trachomatis, Gen-Probe, Inc., San Diego, CA, 1988.
19. **Clyne, J. M., Running, J. A., Stempien, M., Stephens, R. S., Akhavan-Tafti, H., Schaap, A. P., and Urdea, M. S.**, A rapid chemiluminescent DNA hybridization assay for the detection of chlamydia trachomatis, in *Journal of Bioluminescence and Chemiluminescence*, Vol. 4, Pazzagli, M., Cadenas, E., Kricka, L. J., Roda, A., and Stanley, P. E., John Wiley & Sons, Chichester, England, 1989, 357.
20. **Franklin, M. L., Horrlick, G., and Malmstadt, H. V.**, Basic and practical considerations in utilizing photon counting for quantitative spectrochemical methods, *J. Anal. Chem.*, 41, 2, 1969.
21. **Zatzick, M. R.**, How to make every photon count, *Electro-Optical Systems Design*, 20, June 1972.

22. **Malmstadt, H. V., Frankling, M. L., and Horrlick, G.**, Photon counting for spectrophotometry, *Anal. Chem.*, 44, 63A, 1972.

23. **Meade, M. L.**, Instrumentation aspects of photon counting applied to photometry, *J. Phys. Sci. Instrument.*, 14, 909, 1981.

24. **Miska, W. and Geiger, R.**, Luciferin derivatives in bioluminescence-enhanced enzyme immunoassays, in *Journal of Bioluminescence and Chemiluminescence*, Vol. 4, Pazzagli, M., Cadenas, E., Kricka, L. J., Roda, A., and Stanley, P. E., John Wiley & Sons, Chichester, England, 1989, 119.

25. **Bronstein, I., Edwards, B., and Voyta, J. C.**, 1,2-Dioxetanes: novel chemiluminescent enzyme substrates. Applications to immunoassays, in *Journal of Bioluminescence and Chemiluminescence*, Vol. 4, Pazzagli, M., Cadenas, E., Kricka, L. J., Roda, A., and Stanley, P. E., John Wiley & Sons, Chichester, England, 1989, 99.

26. **Schaap, A. P., Sandison, M., and Handley, R. S.**, Chemical and enzymatic triggering of 1,2-dioxetanes. 3: alkaline phosphatase-catalyzed chemiluminescence from an acryl phosphate-substituted dioxetane, *Tetrahedron Lett.*, 28, 1159, 1987.

27. **Kricka, L. J. and DeLuca, M.**, Effect of solvents on the catalytic activity of firefly luciferase, *Arch. Biochem. Biophys.*, 217, 674, 1982.

28. **Hunt, D. T. E. and Wilson, A. L.**, *The Chemical Analysis of Water*, 2nd ed., Royal Society of Chemistry, London, 1986.

29. **Subramani, S. and DeLuca, M.**, *Genetic Engineering*, Vol. 10, Setlow, J. K. and Hollender, A., eds., Plenum, New York, 1988, 75.

30. **Schram, E., Van Esbroeck, H., and Roosens, H.**, Single photon standard for luminescence measurements, in *Proc. 1978 Int. Symp. on Anal. Appl. of Biolumin. Chemilumin.*, Schram, E. and Stanley, P., Eds., State Printing and Publishing, Westlake Village, CA, 1979.

31. **Allen, R. C.**, Phagocyte oxygenation activities: quantitative analysis based on luminescence, in *Bioluminescence and Chemiluminescence: New Perspectives*, Schölmerich, J., Andreesen, R., Kapp, A., Ernst, M., and Woods, W. G., John Wiley & Sons, Chichester, England, 1987, 13.

32. **Pazzagli, M., Cadeneas, E., Kricka, L. J., Roda, A., and Stanley, P. E.**, *Journal of Bioluminescence and Chemiluminescence: Studies and Applications in Biology and Medicine, Proceedings of Vth International Symposium on Bioluminescence and Chemiluminescence*, Vol. 4, No. 1, John Wiley & Sons, Chichester, England, 1988.

INDEX

A

Acetobacter xylinus, 83
Acridine orange direct counting procedure, 131
Aerobiological sampling systems, 112—113
Aerosol experiments, in bioluminescent ATP assay, 115—116
AGI-30 sampler, 113—114, 121
Ammonium sulfate, in immobilization of luciferase, 89—91
Antibiotic susceptibility test (AST), 10
  bioluminescence, 10—12
  Kirby-Bauer method, 11
  rapid results for, 22—23
Apyrase, in hydrolysis of bacterial ATP, 38
Arsenate, stabilization effects of, 88
ATP (adenosine triphosphate). See also Bacterial ATP; Native milk ATP; Somatic cell ATP
  analysis of, 91
  bioluminescent assay of, 113
  in experimental food microbiology, 64
  firefly luciferase assay of, 130
  as index of microbial biomass, 64—65
  quantification and extraction of, 65—67
ATP bioluminescence, in brewing process, 101—103
ATP technology
  estimation of pasteurized milk quality using, 44—46
  estimation of raw milk quality using, 30—31, 34—35
    bacterial ATP, 37—39
    instrumentation and reagents, 40
    native milk ATP, 31
    somatic cell ATP, 35—37
    techniques for, 40—44
Automatic reagent injection, 141—142

B

Bacillus globigii, 114, 116, 118, 121—123, 125, 126
Bacteria. See also specific bacteria
  in brewing process, 98, 99
  detection of, 50—51
    in dairy products, 50—51
    with differential extraction, 54
    with differential growth, 53—54
    lux gene technology, 52—53
  differential extraction of, 54
  extraction of, 37—38
  hydrolysis of, 38
  reaction with luciferase-luciferin, 42
  techniques for enumerating
    extraction and hydrolysis of nonbacterial ATP, 40—41

methods involving concentration of bacteria, 44—45
Bacterial cellulose support, preparation of, 83
Bacteriophage, novel methods to detect, 50
BactoFoss instrument, 69—70
BactoFoss method, 64
  operating procedures in, 71
  on raw meat, 71, 73
    comparison of, 76—77
    homogenization, 73—75
    meat samples, 73
    reference method, 74
    results, 75
    sample treatment in, 72
Bac-T-Screen system
  costs of, 8
  mechanistics of, 5—6
BC-ATP test, 44
Biodeteriogens, bioluminescent, 133—134
Biodeterioration, 130—134
Biodeterioration studies, 130—132
Bioluminescence. See also ATP Bioluminescence
  compared with filtration, 55
  comparison with plate count results, 107
  defined, 2, 130
  detection of milk-degrading enzymes by, 47—48
  to determine ATP, 65
  estimation of bacterial count in raw milk by, 32—33, 43
  in food industry, 67—71
  to monitor starter activity, 48
  properties of, 2—4
  for rapid antibiotic susceptibility testing, 10—11
  reasons for not using, 55
  for UTI, 4—10
  at Wellpark Brewery, 103—108
Bioluminescent ATP assay
  comparison of assay methods, 117
  materials used for, 113
    all-glass impinger for, 113—114
    cyclone (air centrifuge) sampler, 114—115
  methods, 115—116
  results, 117—121, 126
Biotechnology, defined, 112
Bossuyt's method, 43
Brewing process
  ATP bioluminescence in, 101—103, 107
  microbial spoilage in, 98—99
  microscopical methods in, 100—101
  requirements for, 99—100

C

C^{14}, detection of, 101
Cell viability experiments, in bioluminescent ATP assay, 116

Chemiluminescence, 2
Chemstrip LN
  costs of, 8
  mechanistics of, 5
Chlorhexidene, as extractant, 104
Chocolate milk, sterility testing of, 47
Cleaning-in-place (CIP) systems, of modern
    brewery, 98
Concanavalin A, 44
Cost, in evaluation of urine screens, 7—8
Cyclone sampler, 114—115, 118, 120—121, 125

D

Dairy products, detection of bacteria in, 50—51.
    See also Milk
Detectors, in luminescence instrumentation, 140—
    141
Dialysis film, 84
Dynatech, 18

E

Enterobacter cloacae, 16
Escherichia coli, in brewing process, 108, 115,
    118, 123

F

False negatives
  in urine screen, 6—7
  in UTIscreen,® 9
False positives, 6—8
Fentogram (fg) ATP/cfu, ATP content in, 65
Filtration
  vs. bioluminescence, 55
  estimating bacteriological quality of milk by,
    41—42
Fireflies, cost of collecting, 9
Firefly ATP bioluminescent assay, advantages of,
    77
Firefly luciferase. See also Luciferase
  activation of low concentrations of, 90
  activity of immobilized, 85
  coupling to cellulose, 83—86
  immobilization of, 82—83
    effect of diffusion on, 92—93
    preparations for, 87
  reaction kinetics, 88—92
  results with, 86—88
  storage stability of, 88
Fluorescein diacetate (FDA), spectrophotometric de-
    termination of hydrolysis of, 131
Food industry
  bioluminescence in, 67—71
Fresh meat, bioluminescence of, 70

G

Gentamicin, bioluminescence susceptibility test for,
    12

Gram stain
  costs of, 8
  mechanistics of, 5
Glycerol, in immobilization of luciferase, 89—91

H

Hemocytometer counting, in bioluminescent ATP
    assay, 116
Homogenization, with BactoFoss method, 73—74
Hygiene monitoring, 55

I

Immunological methods, for detecting wild yeasts,
    100
Instrumentation. See also LAD system; Luminome-
    ters
  AGI-30 sampler, 113—114, 121
  cyclone sampler, 114—115, 118, 121, 125
  luminescence, 140—147

K

Klebsiella pneumoniae, 15, 54

L

Lactic acid bacteria (LAB) medium, 106
Lactobacilli
  in brewing process, 100—101
  detected in beer, 104
Lactobacillus acidophilus, 49
Lactococcus lactis, 49
Lactococcus thermophilus, 49
LAD model 535 UTIscreen® system, 7—8
LAD Sensi-Quik®
  antibiotic concentration curves for, 14, 17
  current status of, 20
  development of, 12
  error rates for, 14—15, 17
  incubation times for, 14, 16
  ionic strength effects for, 13—14
  medium studies of, 13
  microtiter plate (MTP) luminometer evaluations
    for, 17—21
LAD UTIscreen®
  evaluation of, 6—9
  mechanistics of, 5—6
Lectin-microbe interactions, 44
Lipases, bioluminescence detection of, 48
LN strip, mechanistics of, 5—6
Luciferase, 2. See also Firefly luciferase
  in extraction of ATP, 66
  immobilized, 9
Luciferase-luciferin, reaction of bacterial ATP with,
    42
Luciferase reagents, in brewing process, 103
Luciferin, 2, 9
Luminometers, commercial
  automatic, 144—146
  companies supplying, 147—148

portable, 147
semi-automatic, 146—147
LAD, 7
MTP, 19
Luminoskan, 18
lux gene technology, 52—53, 55

M

Meat, bioluminescence of, 70. See also Raw meat
Media
  development of, 108
  lactic acid bacteria, 106
Membrane filtration, for detecting wild yeasts,
  100—101
Microbiology, bioluminescence in, 23
Microtiter plate luminometer, 17—20, 21
Milk. See also Native milk ATP
  estimate of pasteurized quality in, 45—46
  estimation of raw milk quality, 30—31
    with bacterial ATP, 37
    instrumentation and reagents for, 40
    native milk ATP, 31, 34—35
    somatic cell ATP, 35—37
    techniques for, 40—44
  factors affecting sensitivity of bacterial ATP in,
    34
  microbial ATP in dried, 51
  quenching effects on light emission of, 39—40
Milk Microbial ATP Kit, 55

N

NASA, bioluminescence AST developed by, 11
Native milk ATP, 31—35
Nitrate reductase test (NRS)
  in hydrolysis of bacterial ATP, 38
  mechanistics of, 5—6
Nucleic acid probes, 23
Nucleotide extractants, 54

P

Pasteurized milk, postcontamination in, 69—70
Photo bacteria, 2
Photometers, development of, 108. See also Instru-
  mentation
Plate counts, as reference method, 45—46
Predictive values, evaluation of, 6, 7
Processing plant, hygiene monitoring of, 46
Proteases
  bioluminescence assay for, 55
  bioluminescence detection of, 48
Pseudomonadaceae, in milk, 44
Pyrophosphate, stabilization effects of, 88

Q

Quality control, for luminescence instrumentation,
  143—144

R

Raw meat
  BactoFoss methods on, 73—77
  bioluminescence of, 70
Raw milk, hygienic quality of, 68. See also Milk
Recombinant DNA technology, 112

S

Saccharomyces cerevisae, 115, 117
Salmonella typhimurium, 52
Sensi-Quick™, 10
  antibiotic concentration curves for, 14, 17
  current status of, 20—22
  development of, 12
  error rates for, 14—15, 17
  incubation times for, 14—16
  ionic strength effects for, 13—14
  medium studies of, 13
  MTP luminometer evaluations for, 17—21
Sensitivity, in evaluation of test systems, 6—7
Somatic cell ATP, 35—37
Somex A, 55
Sporeformers, contamination due to, 47
Staphylococcus, two-hour ATP production by, 13
Starter cultures, monitoring activity of, 48—51
Streptococci, lactic acid-producing, 50

T

Test efficiency, 6
Textile challenge testing, 133
Total viable count (TVC) method, vs. BactoFoss
  method, 74—78
Triton® X-100, in hydrolysis of bacterial ATP, 38

U

Ultra heat treatment (UHT) products, sterility testing
  of, 47
Urinary tract infection
  bioluminescence as rapid screen for
    evaluation for, 6—9
    and sociology of UTI, 4—5
  techniques, 5
Urinary tract infection syndromes, 4
UTIscreen™
  comparison of, 6—7
  costs of, 8

V

Vibrio fischeri, 50

W

Wellpark Brewery, 98

bioluminescence at
  chlorhexidine as extractant, 104
  detection of yeast and lactobacilli in beer,
    104—107
  general findings, 103
detection of yeast and lactobacilli in beer
  method, 105—106
  results, 106—107

Y

Yeast, detected in beer, 104—107

Z

Zeta Plus filters, 42
Zeta potentials, of bacteria, 42

Printed and bound by CPI Group (UK) Ltd, Croydon, CR0 4YY

22/10/2024

01777632-0015